Customized Production Through 3D Printing in Cloud Manufacturing

Customized Production Through 3D Printing in Cloud Manufacturing

Lin Zhang

Longfei Zhou

Xiao Luo

ELSEVIER

Elsevier
Radarweg 29, PO Box 211, 1000 AE Amsterdam, Netherlands
The Boulevard, Langford Lane, Kidlington, Oxford OX5 1GB, United Kingdom
50 Hampshire Street, 5th Floor, Cambridge, MA 02139, United States

Notices
Knowledge and best practice in this field are constantly changing. As new research and experience broaden our understanding, changes in research methods, professional practices, or medical treatment may become necessary.

Practitioners and researchers must always rely on their own experience and knowledge in evaluating and using any information, methods, compounds, or experiments described herein. In using such information or methods they should be mindful of their own safety and the safety of others, including parties for whom they have a professional responsibility.

To the fullest extent of the law, neither the Publisher nor the authors, contributors, or editors, assume any liability for any injury and/or damage to persons or property as a matter of products liability, negligence or otherwise, or from any use or operation of any methods, products, instructions, or ideas contained in the material herein.

ISBN: 978-0-12-823501-0

For information on all Elsevier publications visit
our website at https://www.elsevier.com/books-and-journals

Publisher: Matthew Deans
Acquisitions Editor: Brian Guerin
Editorial Project Manager: Czarina Mae S. Osuvos
Production Project Manager: Anitha Sivaraj
Cover Designer: Greg Harris

Typeset by STRAIVE, India

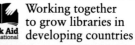

Working together
to grow libraries in
developing countries

www.elsevier.com • www.bookaid.org

Contents

Preface

Since being proposed more than 10 years ago, cloud manufacturing has been integrated with the latest information technology and manufacturing technology. This integration has caused the application scenarios of cloud manufacturing to expand continuously. Changes in the manufacturing industry are becoming more and more significant, and are expected to keep doing so. 3D printing is a technology with a history of more than 100 years. In recent years, this technology has seen fresh developments and has been strongly favored to be one of the core technologies to realize the vision of industry 4.0 due to the pursuit of customized manufacturing. As a special manufacturing technology, 3D printing simplifies many parts of the traditional manufacturing process, and thus can easily achieve the concept of cloud manufacturing. In particular, the ability of cloud manufacturing to support large-scale social manufacturing, combined with the ability of 3D printing to support customized manufacturing, has unique advantages for realizing the long-awaited mass customization.

The idea of combining these two concepts already existed in the early stage of the development of cloud manufacturing, but only a few studies and experiments were conducted in laboratories. This combination was not put into practice on a large scale until it was officially supported by relevant government science and technology plans in 2014. The first 3D printing-oriented cloud manufacturing platform prototype was developed in 2015 by a team led by Lin Zhang. This book summarizes the ideas and experiences of Lin Zhang's team, accumulated over more than 10 years of research and development.

Chapter 1, "Introduction to customized manufacturing," briefly describes the origin and development of customized manufacturing. It presents the changes of the manufacturing industry in customization, production efficiency, production scale, and cost with the progress of production tools from the perspective of the Industrial Revolution. The irreconcilable contradiction between customization and cost under the traditional mode is analyzed. It is suggested that the combination of 3D printing as a subversive manufacturing technology and cloud manufacturing as a revolutionary manufacturing mode will provide an ideal way finally to solve this contradiction.

Chapter 2, "Advances in cloud manufacturing," introduces the basic concept, conceptual model, operation mode, and main characteristics of cloud manufacturing. As a digital, networked, service-oriented, and intelligent manufacturing paradigm, cloud manufacturing gathers manufacturing resources and manufacturing capabilities through a cloud service platform to make full use of sufficient social resources to achieve highly flexible manufacturing.

Chapter 3, "3D printing with cloud manufacturing," discusses the basic working principle of 3D printing, and sets out the specific working mode and benefits of the combination of 3D printing and cloud manufacturing, which is called a cloud 3D printing platform. An architecture of the cloud 3D printing platform and the functions of each layer are described in detail. Several important functional components are briefly introduced, which will be the main subjects of the following chapters.

Chapter 4, "Model design of 3D printing," looks at 3D printing model management and design technology supporting customized printing, and provides a 3D printing model library management framework. On this basis, two 3D model design methods are given. Users can match the sketch of their online drawing with a contour map of the 3D printing model in the model library to obtain the required model for printing. They can also obtain models by uploading multi-angle photos of the product to be printed. Based on a deep learning algorithm of the generative adversarial network, these photos can be automatically transformed into a standard 3D printing model file.

Chapter 5, "3D printing resource access," discusses how to connect a 3D printer to the cloud platform. Connecting a 3D printer to the cloud platform can enable the printer to be accessed remotely and then form 3D printing services to be used by distributed customers; this is an important feature of cloud 3D platform based on cloud manufacturing. This chapter introduces two 3D printer accessing methods: adapter-based and sensor-based. In the adapter-based accessing method, through the communication module of the printer, different adapters can be designed for access to collect the basic information that the printer can provide. In order to conduct more comprehensive monitoring and health management of the printer, additional sensors need to be used to collect the information that the printer cannot provide and the environment information related to the operation of the printer. The sensor-based accessing method was developed for this purpose.

Chapter 6, "3D printing process monitoring," discusses printing process online monitoring based on sensing data and machine learning. Printing failure often occurs due to various faults in the printing process. Two common printing faults are curling in the initial stage of printing and material breaking during printing. By using the sensor-based access technology given in Chapter 5, appropriate sensors are selected to monitor the real-time running status of 3D printers. Based on the data collected from sensors, the machine learning method is used to identify printing faults as early as possible. The proposed online monitoring method can thus reduce the loss caused by equipment failure.

Chapter 7 focuses on "3D printing credibility evaluation." Design and printing are the two most important parts in the production process of 3D printing. The results of design are the models for 3D printing, and the results of printing are the products that users ultimately need. The two activities directly affect whether the printed products meet the requirements of users. Therefore, their credibility needs to be evaluated. This chapter provides credibility indexes or indicators and evaluation methods for 3D models and printing services, respectively.

Chapter 8, "Supply-demand matching and task scheduling," studies the problem of supply-demand matching and task scheduling in 3D printing platforms. For a large number of user orders and a large number of printing services on the platform, it is necessary to find quickly and accurately the most suitable printing equipment for each printing task. This chapter introduces a supply-demand matching method based on multisource data, which provided descriptions of print task and print service, capability indicator model, and matching rules. Then, using an optimization algorithm, accurate matching and scheduling of printing tasks are studied to solve the problems involved when a task has multiple optional printing services and a printing service is matched by multiple tasks.

Chapter 9, "3D printing process management," studies how to improve the efficiency of printing by effective management of the printing process. When a product is composed of multiple parts, multiple printers are required to complete the whole product. This chapter presents a parallel processing service composition optimization method to improve printing efficiency. In addition, a printing task packaging solution is proposed. According to the shape, material, color, and other characteristics of printing products, multiple printing models are packaged and sent to one printer to complete the printing; this can reduce the time spent on auxiliary processes, and thus improve printing efficiency.

Chapter 10, "Security and privacy in cloud 3D printing," discusses security issues related to cloud 3D printing, including data security of cloud 3D printing platforms, access control for cloud 3D printers, the security of cloud 3D printing based on blockchain, and the intellectual property protection problem in cloud 3D printing. Regarding data security, we consider the main data security problems and solutions. Access control methods of cloud 3D printing equipment are also discussed. The chapter then reviews blockchain-based service security technologies and the intellectual property protection problem in the cloud 3D printing environment.

Chapter 11, "Application demonstration of cloud 3D printing platform," introduces a cloud 3D printing platform prototype developed by the authors' team based on the methods and technologies introduced in this book. The customized production process and the main functions of this platform are illustrated through the production process of a storage box.

Chapter 12, "Conclusions and future work," summarizes the significance of the combination of 3D printing and cloud manufacturing for realizing customized mass production with high efficiency and low cost. It also sets out the focus and development directions of cloud 3D printing platform for the future.

The book is organized based on the research results of Lin Zhang and his students during the past decade. These students include Jingeng Mai (Chapters 3, 5, 7, and 9), Longfei Zhou (Chapters 2, 8, and 10), Xiao Luo (Chapters 7, 8, and 11), Fan Pan (Chapter 4), Zemin Li (Chapter 4), Ce Shi (Chapter 5), Bing Li (Chapter 6), Jin Cui (Chapter 7), Zhen Zhao (Chapter 9), and Yuankai Zhang (Chapter 10). Xiaohan Wang helped with drawing and editing some figures.

Finally, we would like to thank Elsevier for providing this opportunity to share our research with peers all over the world. We hope that the ideas and technologies in this book can provide some inspiration for the development of the manufacturing industry and that we may receive feedback from readers to help us conduct better research and applications in the future.

Chapter 1

Introduction to customized manufacturing

1.1 The origin of customized manufacturing

The meaning of the word "Customize" is custom-made, made to order of goods, it can be traced back to the word "Bespoken" (the past tense from bespeak). Now it is usually used in tailored suits. Fig. 1.1 is Savile Row located in Mayfair shopping district, London, UK. It is regarded as a pilgrimage site for custom tailoring, where the term "Bespoken" is considered the most advanced expression of the User's customized demands for tailoring [1]. The service of personal tailoring is the next best thing, which is to make some detailed adjustments based on the existing clothing types according to user requirements. Ready-to-wear represents the service to buy the clothing templates for existing designs. Therefore, customization can be regarded as the service process of providing personalized products for each customer.

With the rapid development of manufacturing tools in the clothing industry, the traditional clothing production mode has been challenged. For the lower-level mentioned above ready-to-wear production mode, the traditional production process generally allocates order tasks according to the process several months in advance. One order may require the production of tens of thousands of the same clothing. After receiving the order, the factories responsible for different processing tasks will conduct large-scale assembly line operations to reduce production costs. However, with the changes in the apparel industry's individual needs, even for ready-to-wear services, sample orders no longer require mass production, generally only a few hundred pieces of production scale.

For most clothing manufacturers, they are usually unwilling to accept small batch orders due to cost considerations. However, as the production of multiple varieties and small batches has become the norm, the continuous pursuit of mass production mode no longer meets current development requirements. Small-batch production has also brought severe challenges to the innovation of feeding, production, inspection and testing, and garment manufacturers' logistics.

Customized Production Through 3D Printing in Cloud Manufacturing
https://doi.org/10.1016/B978-0-12-823501-0.00007-9

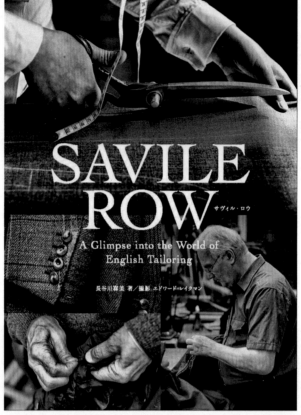

FIG. 1.1   Savile Row—a custom suits pilgrimage site, located in Mayfair shopping district, London, United Kingdom. *(Source: Yosemite Hasegawa.)*

Therefore, those clothing manufacturers that cannot innovate production processes in-time will face the risk of bankruptcy.

Regarding the customization demands of Bespoken-level user groups, the biggest challenge to the traditional clothing industry is to solve data integration throughout the life cycle of the production process. Due to the diversification of customized styles and the individualization of sizes, the production process will be more detailed to realize the one-person-one-version production model truly. At this time, a suit will be produced in a flexible combination with sleeves, pockets, and buttons as the basic units. However, nowadays, under the outsourcing-oriented clothing industry's world pattern, the refined production technology division will inevitably lead to the long delivery integration cycle caused by complicated outsourcing links.

For example, in the 1960s, only 4% of American sportswear was outsourced. By 2014, nearly 95% of sportswear companies relied on outsourcing to produce.

To take advantage of Asia's cheap labor, these sportswear companies have only a small number of internal employees responsible for the design, research and development, and marketing, and the production process is outsourced. Even if it is to produce non-customized sportswear, it usually takes up to 18 months from design to production. In other words, the new clothing that users can buy on the market was designed 2 years ago.

The clothing industry and the entire manufacturing industry are currently facing changes triggered by manufacturing tools' generational development. As mentioned above, the integration of data from the supply chain to the entire product life cycle has always been the driving force behind manufacturing tools' development. Let's imagine the future industrial production, and product customization will become the norm. As the data flow of the product life cycle is integrated, as long as the product production process has not been completed, it can be adjusted flexibly at any time. The optimal allocation of resources will maximize production efficiency and minimize waste generated during the production process. It is exactly the vision of large-scale personalized, customized products in the future manufacturing industry described by Industrial Revolution 4.0.

1.2   Evolution of customized production

From a historical point of view, taking the previous industrial revolutions as the time axis coordinate, from the perspective of the relationship between customized production and production efficiency and cost, we can divide the production mode of goods into four stages, namely, cottage industry production, machine production, mass production, and future production. Now let's give an overview of the evolution of the relationship.

1.2.1   Cottage industry production

Before the first industrial revolution, cottage industry has always been the mainstream production mode and has been passed down to this day. Products were made in manual workshops with very low efficiency. This mode of production is characterized by low efficiency and small output, but high degree of personalization. From a modern perspective, many products are works of art. Lot of valuable craftsmanship have created in long-term practice. For example, the structure of mortise and tenon joint was invented by Chinese craftsman Lu Ban, which can be simple and strong to connect pieces of wood at an angle of 90 degrees. It has been used for thousands of years by woodworkers around the world to join pieces of wood. This structure is called dragonfly when applied to housing construction, and although each component is thin, it can withstand great pressure as a whole. In ancient and modern buildings designed by Chinese and Japanese architects, this structure is widely used, shown in Figs. 1.2 and 1.3.

FIG. 1.2 Teng Wang Ge was built in AD 653, as a landmark wooden building of the structure, can withstand the magnitude eight earthquake, after many earthquakes are still intact standing in Nanchang City, Jiangxi Province, China.

FIG. 1.3 Tamedia was designed by Japanese architect Shigeru Ban, the interior is made entirely of spruce wood, which is stable, beautiful and water resistant. The building is a modern and improved version of the structure of mortise and tenon joint. Since 2000, Tamedia is located in Zurich, Switzerland.

These carpentry skills have been passed down to this day, mainly because the manufacturing model (design drawings) are well preserved.

The Black Forest Cuckoo Clock (German: Schwarzwälder Kucksuhren) is produced in the famous German Black Forest region of southwestern Germany on the border with France and Switzerland [2]. These clocks, every half-point and the whole point, the small wooden door above the clock will automatically open, and there will be a timely cuckoo, a pleasant "grunt" (boo valley) cry, known as the most famous handicrafts, shown in Fig. 1.4. As a representative of the regional family handicrafts, the Black Forest region residents have made low-cost, beautifully shaped cuckoo clocks made from local wood according to local conditions. Each detail bears the essence of the hundred years mentioned above of skill, and the bell's famousness dramatically enhances the quality of life of the local people.

But unfortunately, whether in the East or the West, the craftsmanship of cottage industry has always been regarded as the secrets of a family or a very small

FIG. 1.4 The second-generation Black Forest cuckoo clock, based on houses and living scenes, is not only fine craftwork of art but also full of the passion of artisans for life, while realistic landscape wood carvings contain more of the feelings of artisans for the land of flowers, grasses, and birds.

group. Due to the lack of technology sharing and inheritance mechanism, many excellent skills have been lost. For example, the construction craftsmanship for Egypt's pyramids and the tomb of the First Emperor of China was lost with these builders' burial. The smelting process of the bronze sword, as shown in Fig. 1.5, was also permanently forgotten with the death of the family of the swordsmith. Such as the transition from alchemy to modern chemistry, the inheritance of skills is not about tips and tricks but how to use scientific theories and methods.

The cottage industry can be understood as the one-piece (no batch) and highly personalized production mode. The complicated craftsmanship makes the products difficult to mass-produce. Even if the same product is produced, it is inevitable that the details are different. The product's quality depends entirely on the producer's technical level.

1.2.2 Machine production

The first industrial revolution heralds a new era for machines to replace manual labor. It is also a revolution in the mode of production. With the boom in cross-regional business activity, new consumption opportunities have been spawned.

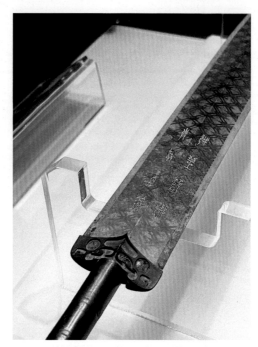

FIG. 1.5 Sword of Goujian, the King of Yue State's bronze sword in the Spring and Autumn Period of China. This sword is now in the Hubei Provincial Museum, Wuhan, China.

Most citizens have greater access to goods and fashionable new items from different regions. For example, since the late seventeenth century, associations and clubs have popped up, and Cafes have sprung up in the streets of France and London. Social activities drive consumption across classes, and lending is common in everyday life. Under such social background, the cottage industry production mode, with the family workshop as the basic unit, can no longer meet the population's material needs.

With the invention of new machines such as foot sewing machines and power looms, production efficiency and production scale have been greatly improved, and the factory system has become a popular form of production organization. It brings together people engaged in productive activities so that they can share production tools or energy. The original family workshop-style self-sufficient production mode was broken, the fate of the craftsman was also involved in the torrent of the times. When a large number of men, women and children use bloodless machines for production, the title of craftsmen is replaced by workers. The speed of factory production is no longer determined by the individual but by the production manager. Machine production is gradually replacing the skills of craftsmen and becoming the main method of industrial production. With the help of machines, the efficiency of production and the degree of standardization of products were greatly improved, meanwhile, the degree of personalization is significantly reduced.

1.2.3 Mass production

With the advent of the second industrial revolution, interchangeability and production lines have become the core development concepts of this era,. It can further reduce the heavy physical labor of workers and improve production efficiency. To put it simply, interchangeability means that each component's production error must be within an acceptable range. Thanks to this, the parts produced by different types of machines can be universal to a certain extent so that the various parts of the product can be mass-produced and assembled at will.

At the beginning of the 20th century, Ford established a factory in Dearborn, Michigan, United States. This factory was the first enterprise to introduce a "modern" production line, which opened the era of mass production. The modern production line is also the assembly line. With the help of the improved assembly line technology of the conveyor belt, workers can stay in place on their workbenches and be responsible for more refined individual process tasks.

For example, there are approximately 84 processes on the assembly line of the Ford Model T. The assembly line has shortened the assembly time of a car from 12 h to 90 min and increased productivity by eight times. They have created unprecedented economies of scale, reducing the cost of automobile production to an incredible level. Such a huge change has made the car no longer a consumer product for the wealthy but a means of transportation for the masses. From 1908 to 1927, 15 million vehicles were sold, accounting for more than 56% of the American automobile market at that time. More than half of the cars on American highways were Ford Model T cars, as shown in Fig. 1.6.

Although the second industrial revolution greatly improved machines' productivity and reduced the cost of products, but they lost their individuality. The products on a production line were manufactured according to unified standards. Different products almost had no difference in look and use. Customers had no choice. The degree of personalization reached the lowest level.

1.2.4 Mass customized production

With the increasing capacity of mass production and more and more abundant products, consumers are no longer satisfied with having the same product as others. They begin to pursue being different, which poses a greater challenge to enterprises.

Because it is impossible for enterprises to return to the original cottage industry production or machine production to meet the personalized needs of each product, it needs an advantage that can not only meet the personalized needs of products, but also give full play to the efficiency and cost of large-scale

FIG. 1.6 The Ford Model T was introduced to the market in 1908, and the price was set at US$825, which was only about 1 year's salary for a teacher.

production, which gave birth to the mass customization model. In 1970, American futurist Alvin Toffler put forward a new idea of production mode in his book Future Shock [3]. That was to provide products and services with specific needs of customers at a cost and time similar to standardization and mass production. In 1987, Stanley M. Davis first named this production method as Mass Customization (MC) in his book Future Perfect [4]. In 1993, B. Joseph Pine II wrote in the book Mass Customization: The New Frontier in Business Competition [5]: "At its core, is a tremendous increase in variety and customization without a corresponding increase in costs. At its limit, it is the mass production of individually customized goods and services. At its best, it provides strategic advantage and economic value."

In order to improve the customization ability of production process, a series of methods and technologies have been developed, such as flexible manufacturing system (FMS), modular design and manufacturing, agile manufacturing (AM) and so on. With the advent of the third industrial revolution, digital technology plays an increasingly important role in manufacturing, and also provides strong support for the realization of the goal of mass customized production.

Innovations supported by digital technologies are used to improved the productivity and reduced the cost of customized production. For example, Coca-Cola, which is well-known for standardization, is also trying to start mass customization. With the help of Hewlett-Packard's new printing technology, it provides millions of unique label customized services for different countries' special needs. This new customized label service has greatly increased Coca-Cola's sales and surpassed the total sale volume of the classic 20-oz plastic bottles produced by standardization.

The globalization of manufacturing also provides a favorable environment for reducing the cost of personalized manufacturing. By the 1990s, manufacturing globalization had reached an unprecedented level. The concept of the global value chain was born [6]. The global value chain refers to all the necessary activities that a product experiences, from the design concept's formation to the consumer's final arrival. The product design, production, marketing, distribution, after-sales service, and support are all included. In the global value chain, in order to optimize the resource allocation of enterprises, well-known enterprises in the manufacturing industry control product design and product production, and many production tasks, including product assembly, are entrusted to suppliers to complete. Global manufacturing enables enterprises to make use of more manufacturing resources and manufacturing capabilities worldwide in order to minimize costs.

The progress of manufacturing equipment also provides support for mass customization. Obsolete machines without intelligence are more in line with standardized large-scale mass production models. Under the support of communication, computer, and modeling and simulation technology, those cold machines have been given new life-like characteristics by technology. Through the embedding of various sensors, data in the machine production process can be effectively obtained. Driven by the wave of digital production, more software defined smart machines and modularized composable production lines are favored by the entire industry. At the same time, smart machines have also promoted the digital development of the entire industry chain. Digital and modularized production makes products no longer limited to the use of traditional crafts or machines. The entire production line is more like an organic life body [7], and intelligent production equipment can perceive the surrounding production environment and the execution status of related equipment, thereby increasing the flexibility of production.

These efforts based on the improvement of traditional manufacturing technologies and equipment were all attempts to get a better tradeoff between customization and production efficiency (cost). However, the degree that could be achieved was still limited, which was far from meeting the needs of the market.

1.2.5 Future production

With the emergence of a batch of new generation information technologies such as cloud computing, big data, Internet of things, cyber physical system, and artificial intelligence, as well as the rise of a batch of new manufacturing modes and technologies, such as Industrial Internet of Things, smart factory, 3D printing and cloud manufacturing, the manufacturing industry has ushered in a new development climax and is setting off a new era of new revolution characterized by intelligent, networked, service-oriented and customized, which is knowed as the fourth industrial revolution. Through this revolution, customization, production efficiency and cost will achieve a perfect balance in the future.

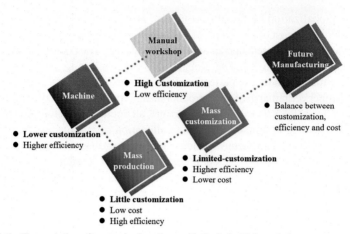

FIG. 1.7 The evolution of customization along with the industrial revolutions.

In conclusion, the evolution process of the relationship between customized production and production efficiency and cost can be summarized as follows (Fig. 1.7). At the beginning of the manufacturing industry, such as the period of cottage industry production, everything was made manually. The customization was at the highest level, but the efficiency of production was very low. When humans went into the first industrial revolution, things can be made by machines. The efficiency of manufacturing was greatly improved. But the customization level became lower. In the early 20th century, the first production line was invented, manufacturing entered a new mass production era. The production efficiency was greatly improved, and the cost was significantly decreased. While the customization reached the lowest level. In 1980s, the concept of mass customization was proposed, but because of the limitations of technologies, the customization level can only be improved very limitedly. Now we are on the way to a revolutionary changes in manufacturing, which will completely solve the contradiction between customization and production efficiency and cost.

1.3 3D printing and cloud manufacturing

It can be seen from the analysis in the previous section that under the traditional manufacturing mode, the increase of production efficiency often leads to the decrease of cost, but at the same time, the degree of customization will also be reduced. The contradiction between efficiency (cost) and personalization seems irreconcilable. One of the most important objectives of the next generation manufacturing is to solve this contradiction with the support of many new information and manufacturing technologies. In particular, the emergence of 3D printing technology has greatly increased the possibility of achieving this goal.

3D printing is also known as additive manufacturing technology, which produces parts by stacking layer-by-layer manufacturing concepts. Therefore, objects of any shape can be printed. Its production process is born with the genes of customized and digital manufacturing. 3D printing has been expected to bring a new industrial revolution [8]. At present, 3D printing has begun to play a role in some manufacturing fields. For example, the garment industry mentioned at the beginning of this chapter shows high enthusiasm in using 3D technology.

In recent years, some clothing manufacturers have turned their attention to 3D Printing to reduce the clothing industry's production cycle. By adopting whole shaped technology, the integration cycle of the product is greatly reduced. For instance, in the speed factor of Adidas [9], 3D printing simplifies the previously scattered and cumbersome product process, the simple process of sole stitching is usually completed by multiple outsourcing companies such as shoe uppers, shoe soles, shoelaces, etc. through multiple different processes, which greatly increases the turnover cycle of the supply chain. After introducing 3D Printing whole shaped technology, the above process can be completed in just a few hours, as shown in Fig. 1.8. Under the supply chain innovation triggered by 3D printing technology, other major sports brands are also actively acting. Nike has teamed up with its suppliers to create a "Manufacturing Revolution," which can double the production speed [10]. Under Armor, an American sports clothing has established a technology incubation project called the lighthouse. The project integrates 3D printers, individualized foot scanners, and automated assembly robots to accelerate the design and production of new products.

As a pioneer in applying 3D Printing manufacturing technology in the automotive industry, BMW has been incorporating 3D printed parts into the concept car research and development system since 1991. In 2010, BMW first began to

FIG. 1.8 3D Printing shoes—Adidas Futurecraft 4D "Ash Green" offer price 300$.

FIG. 1.9 Left: the Individual Night Sky of BMW. Right: brake calipers, the parts produced by 3D printing technology.

use plastic and metal-based process technology, which was initially mainly used to produce smaller parts. At the 2019 Geneva Motor Show, the Individual Night Sky of BMW (see Fig. 1.9) is a foolproof glamorous look for everyone. The meteorite debris material used in the cabin and the unique space theme rendered the car mysterious and attractive. Part of the exterior materials made by 3D printing technology can reduce the weight by 30% while achieving more detailed and intricate patterns, significantly improving driving dynamics and riding comfort. It is reported that with the help of 3D printing technology, the guide rail was developed in only 5 days, and mass production was soon carried out at the Leipzig factory.

In the aviation manufacturing industry, 3D printing technology has also begun to be valued and applied. Under the support of 3D printing technology, customized replacement production of complex products has been realized. ATP type of aircraft engine designed and built by © GENERAL ELECTRIC through 3D Printing [11], as shown in Fig. 1.10. The number of components needed for an entire aircraft engine has been reduced from 855 to just 12. 3D printing technology can significantly reduce the potential risk caused by multi-components welding.

FIG. 1.10 GE announced that the ATP aircraft engine currently require only 12 components by 3D Printing. *(https://www.ge.com/reports/mad-props-3d-printed-airplane-engine-will-run-year/.)*

However, although 3D printing simplifies the traditional manufacturing process and naturally has the characteristics of personalized production, it is difficult to fundamentally improve the manufacturing efficiency if large-scale production cannot be formed.

Cloud manufacturing is a new networked, service-oriented and intelligent manufacturing mode that concentrates a large number of manufacturing resources and capabilities on the cloud service platform [12,13]. With the help of cloud manufacturing service platform, a flexible supply chain can be built by making full use of the rich manufacturing resources and capabilities of the whole society, so as to provide strong support for large-scale, efficient and low-cost customized manufacturing. The combination of 3D printing and cloud manufacturing will bring customized mass production to a new level.

1.4 Conclusion

In this chapter, firstly, we use the etymology study of the word "Customize" in the clothing industry to introduce the changes that the development of production tools has brought to customized manufacturing in the clothing industry. Secondly, along the process of the industrial revolution, this chapter analyzes the development process of customized manufacturing and the changes of the relationship between the degree of customization, efficiency and cost in this process. Under the traditional manufacturing mode, there is an irreconcilable contradiction between them, but with the development of new generation information technology and the emergence of new manufacturing technology, it is full of hope to solve this contradiction. The combination of 3D printing and cloud manufacturing will be an ideal way to finally realize efficient and low-cost personalized manufacturing.

References

[1] Savile Row Bespoke, Championing half a Century of Sartorial Excellence. https://www.savilerowbespoke.com.

[2] House of 1000 Clocks, Cuckoo Clocks, Originally and Traditionally Made in the Black Forest. https://www.hausder1000uhren.de/.

[3] A. Toffler, Future Shock, Random House, New York, 1970.

[4] S.M. Davis, Future Perfect, Addison Wesley, 1987.

[5] B.J. Pine, Mass Customization: The New Frontier in Business Competition, Harvard Business Review Press, 1992.

[6] G. Gereffi, The organisation of buyer-driven global commodity chains: How US retailers shape overseas production networks, in: G. Gereffi, M. Korzeniewicz (Eds.), Commodity Chains and Global Capitalism, Praeger, Westport, CT, 1994.

[7] C.B. Frey, The Technology Trap: Capital, Labor, and Power in the Age of Automation, Princeton University Press, 2019.

[8] I. Whadcock, The third industrial revolution, De Economist (2012). Apr. 21.

[9] R. Manthorpe, Adidas's secret weapon, Wired (2017). Nov. 2.

[10] Nike, Nike's Manufacturing Revolution Accelerated by New Partnership with Flex. https://news.nike.com/news/.

[11] V.D. Matthew, GE's 3D-Printed Airplane Engine Will Run this Year, 2017.

[12] B.H. Li, L. Zhang, S.L. Wang, et al., Cloud manufacturing: a new service-oriented networked manufacturing model, Comput. Integr. Manuf. Syst. (in Chinese) 16 (1) (2010) 1–8.

[13] L. Zhang, Y.L. Luo, F. Tao, et al., Cloud manufacturing: a new manufacturing paradigm, Enterp. Inf. Syst. 8 (2) (2014) 167–187.

Chapter 2

Advances in cloud manufacturing

2.1   A new paradigm of manufacturing

The concepts of cloud manufacturing was first proposed in 2009 [1,2]. Cloud manufacturing is a new service-oriented, efficient and low consumption networked intelligent manufacturing model. It integrates existing technologies including information manufacturing, cloud computing, Internet of Things, service computing, intelligent science and high-performance computing, extends and changes traditional networked manufacturing and service technologies, to realize intelligent, win-win, universal and efficient sharing and collaboration by making all kinds of manufacturing resources and capabilities visualized and service-oriented and carring out unified and centralized intelligent management of the services. Cloud manufacturing can provide high-quality, cheap, safe and reliable services that can be obtained at any time and used on demand through networks for the whole life cycle process of manufacturing, which will make manufacturing resources and capabilities as easy to use as water, electricity and gas. The idea of cloud manufacturing has been widely accepted by academia and industry worldwide in the past decade.

Keeping on the idea of "centralized use of distributed resources and integrated management of centralized services," a cloud service platform can be built to provide a brand new industrial ecological environment to help manufacturing enterprises at different development stages of industry 2.0, 3.0, or 4.0 to improve their innovation ability, customized manufacturing ability and market competitiveness.

Compared with traditional manufacturing enterprises' closed environments, the cloud manufacturing service platform is committed to achieving a cross-enterprise and cross-sector open manufacturing environment. The information of the whole supply chain can be fully shared. A virtual enterprise with huge manufacturing ability and high flexibility can be formed with the support of the cloud platform to efficiently finish orders placed on the platform.

The order-demand-oriented manufacturing mode will blur and eliminate the boundaries between internal and external manufacturing companies, forming a

Customized Production Through 3D Printing in Cloud Manufacturing
https://doi.org/10.1016/B978-0-12-823501-0.00008-0

larger ecosystem of mutual benefit among these companies. In this ecosystem, user-centered and service-oriented characteristics are the two crucial culture signs. In terms of the user-centered characteristic, to achieve customized manufacturing driven by user demand, the most crucial thing is to eliminate the natural barrier between user and production process. With the support of Model Based System Engineering (MBSE), Cyber Physical System (CPS), Internet of Things (IoT), Digital-twin, and other sophisticated technologies, the barrier can be gradually eliminated. These new generations of advanced technologies bring users closer to the entire production process and provide users with a more realistic and immersive experience. And more importantly, while improving the users' experience, it consolidates customer relationships and obtains a large amount of first-hand user feedback data. In terms of service-oriented characteristic, the cloud platform revolutionizes the mutual benefit model of stakeholders. It can provide a reciprocal policy for different parties from hard resources, soft resources, and human resources. In brief, the manufacturers with manufacturing equipment can provide hard manufacturing resources to the platform in a leased manner. The teams with software development qualifications can develop the appropriate service components that can be provided as soft resources of the platform. The professionals with a special design or operational skills can join the platform as human resources.

2.2 The concept of cloud manufacturing

As mentioned in the previous section, cloud manufacturing is a new manufacturing paradigm. It integrates manufacturing resources and abilities via virtualization and service technology as a cloud pool. It manages and operates them uniformly to support intelligent, universal, and efficient sharing and collaboration, to provide safe, high-quality services on demand through networks the whole life cycle of manufacturing.

- Combine the existing manufacturing information technologies (information design, production, experiment, simulation, management, integration) with hot/new information technologies (e.g., cloud computing, the internet of things/CPS, service science, intelligent science, high-performance computing, etc.)
- Integrate manufacturing resources and capabilities via virtualization and servitization as a cloud pool, and then manage and operate them uniformly to support intelligent, universal, and efficient sharing and collaboration.
- Provide safe, high-quality services on demand through networks for the whole life cycle of manufacturing.

The cloud service platform based on the concept of cloud manufacturing is dedicated to enhancing the user's immersive experience, enabling users to participate in the manufacturing life cycle activities through related cloud services. A brief introduction to cloud manufacturing is given as follows.

2.2.1 The conceptual model of cloud manufacturing

The conceptual model of cloud manufacturing can be abstracted as three parts, three roles, two processes, and one support [2,3], as shown in Fig. 2.1.

Three parts: include manufacturing resources and capabilities, applications in the whole lifecycle of manufacturing, and manufacturing cloud.

- Manufacturing resources usually refers to equipment resources that can be connected to the cloud platform and remotely operated and controlled. Manufacturing capabilities includes human resources, professional knowledge, and hardware resources that cannot be directly connected to the platform, which are usually presented in the form of digital models or representations.
- Applications in the whole lifecycle of manufacturing include all kinds of activities of users in different phases of the whole manufacturing lifecycle, such as design, simulation, production, management, purchase, sales, maintenance.
- Manufacturing cloud refers the cloud service platform composed of a large number of manufacturing services and various of tools to support activities of different roles in the platform.

Three roles: include resource/capability providers, resource/capability users, and cloud operator.

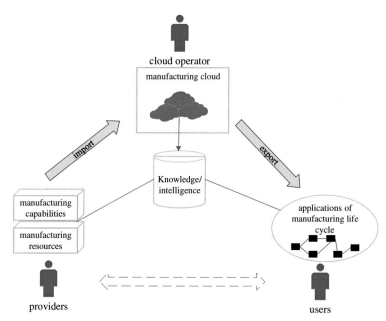

FIG. 2.1 The conceptual model of cloud manufacturing [2].

- Cloud service platform realizes management of manufacturing resources/ capabilities and maintenance of applications through three roles defined in the conceptual model of cloud manufacturing. The three roles include cloud operator, resource provider, resource users. Among them, the **providers** provide the manufacturing resources and/or capabilities to the platform. The provider carries out daily maintenance and management of the provided resources/capabilities. Simultaneously, the platform establishes resources/capabilities specifications to actualize unified access, registration, and use of the resources/capabilities. The **cloud operator's** responsibility is to operate and maintain the platform to provide high quality services to both providers and users. Users include all kinds of **users** in the whole lifecycle of manufacturing. The roles of providers and users are interchangeable.

Two processes: include import and export.

- Import: to connect or map the soft or hard manufacturing resources and capabilities, encapsulate them into services, and store and manage them in the cloud.
- Export: to get the manufacturing services from the manufacturing cloud through networks by different manufacturing users. Users can use services in the on-demand way or collaborate with each others with the supports of application tools in the cloud.

One support: data, models and knowledge gathered in the cloud platform, and AI technologies give strong support to operations and activities in the platform to make them intelligent.

2.2.2 Whole life cycle activities in a cloud environment

The manufacturing lifecycle activities can be conducted or managed in the manufacturing cloud platform, including design, simulation, production, assembling, logistics, planning and scheduling, after-sales service and maintenance. A manufacturing task can be divided into different activities. The cloud platform will find the most suitable services based on the activities to form a manufacturing process or a virtual enterprise. The collaboration of these services can finish the manufacturing task, and the product will be delivered to the user through logistics [2,4], as shown in Fig. 2.2.

From the perspective of customized manufacturing, every step of the product lifecycle is presented to the user as a service with the support of enabling technologies, such as cloud computing and container technology. A service can be defined as a functional component arranged in the workflow of customizing orders for users, providing necessary data support for related nodes of the workflow. The service components will be centrally and uniformly managed in the platform's service pool. Finally, the platform will provide users with the

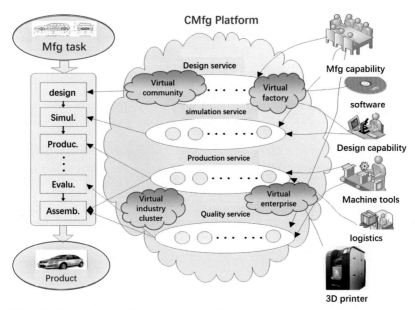

FIG. 2.2   How a cloud manufacturing system works.

most reliable service combination solutions. Users can use customized softwares to view their products' processing status at different stages and give real-time feedback to the platform (Fig. 2.3).

2.2.3   Six unique abilities of cloud service platform

For a manufacturing cloud platform, there are six unique abilities [3,5]:

(1) Decentralized Resource Gathering and Optimization: With virtualization and service-oriented technologies, manufacturing resources and capabilities are gathered to form an enormous resource pool that can be expanded indefinitely;

(2) On-demand Use of Manufacturing Services: Providing clients with manufacturing resources and capabilities in the form of manufacturing services through the network at all times and places; supporting free trading, circulation and on-demand use of manufacturing resources and capabilities;

(3) Data and Knowledge Gathering and Sharing: Gathering data, models, experiences, and knowledge involved in manufacturing to support manufacturing innovation;

(4) Social Manufacturing: Bringing together small and micro-sized enterprises through dynamic alliances and crowdsourcing based on the cloud service platform. An enterprise itself may be small, but the unification of many such enterprises will create sizeable manufacturing capabilities;

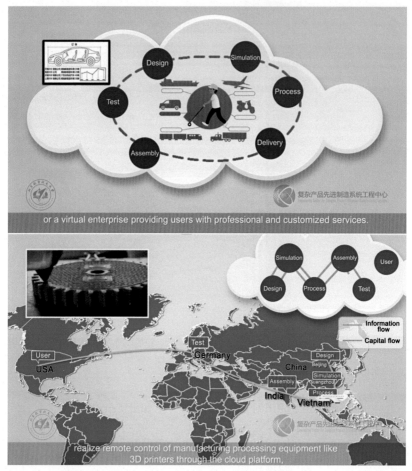

FIG. 2.3 A virtual enterprise based on a cloud manufacturing platform, which combines CPS, IoT, V.R. technologies.

(5) Highly Flexible Manufacturing: Forming a highly flexible virtual compliance production line through the intelligent combination of componential manufacturing services, achieving highly personalized manufacturing capabilities;

(6) Highly Personalized Manufacturing: Provision of personalized and customized services based on the capabilities stated above. In particular, the combination of cloud manufacturing and 3D printing provides strong support for truly individualized manufacturing.

These abilities can significantly improve the level of manufacturing resource sharing and collaboration, improve innovation ability based on the crowd intelligence, especially for small and medium enterprises, and, most importantly, improve the level of customization of manufacturing.

2.3   Six technical features of cloud manufacturing

For a cloud manufacturing system, there are six technical features to support the new model of cloud manufacturing, including the characteristics of "digitization, networking, virtualization, service-orientation, collaboration, and intelligence" of manufacturing resources and capabilities. These six technical features are interconnected and progressive. They are important symbols that distinguishes cloud manufacturing from other information-based manufacturing systems.

2.3.1   Digitization

The digitization of cloud manufacturing systems refers to the transformation of information on manufacturing resources and manufacturing capabilities (including static attributes, dynamic behaviors, etc.) into numbers, data, and models for unified processing including analysis, planning, reorganization, and manipulation. Through digitization, manufacturing resources and capabilities are transformed into manufacturing resource/capability systems that can be monitored, controlled, and managed, such as CNC machine tools, robots, computer-aided design software, management software, etc. The "digital" objects of cloud manufacturing systems include manufacturing resources and capabilities in the entire life cycle such as product design, simulation, production, testing, and operation management. The digitization of manufacturing resources is the basic technology of manufacturing informatization, and it is also the premise and foundation of cloud manufacturing.

2.3.2   Networking

The integration and optimization of people, organizations, and technologies in the whole life cycle of manufacturing is the core to realize advanced manufacturing mode. Therefore, cloud manufacturing systems integrate information technologies such as the Internet of Things and cyber-physical systems (CPS). Using these technologies can realize the thorough access and perception of soft and hard manufacturing resources and capabilities in the whole system. Cloud manufacturing focuses more on the access and perception of hard manufacturing resources (such as machine tools, machining centers, simulation equipment, test equipment, and logistics) and hard manufacturing capabilities (such as people and organizations). Under the cloud manufacturing mode, various soft and hard manufacturing resources can be automatically or semi-automatically accessed through corresponding adapters, sensors, barcodes, RFID, cameras, human-machine interfaces and other devices and technologies. Based on collecting the status of various soft and hard manufacturing resources, the information is transmitted to the cloud manufacturing platform using the communication network to serve business execution processes in cloud manufacturing. For example, based on the perception and analysis of the temperature, pressure, load, and other information of chemical reaction devices, it provides a basis for production planning and task scheduling. The real-time tracking of logistics goods can assist virtual enterprises to monitor and manage the transaction execution process of upstream and downstream partners.

2.3.3 Virtualization

Virtualization technology originates from the research on virtual machines in the computing field. Virtualization is the core technology of cloud computing platforms. Manufacturing resource virtualization refers to the abstract representation and management of manufacturing resources. It is not bound by specific physical limitations. To achieve virtualization, we need to provide standard interfaces for inputs and outputs of manufacturing resources. The objects of manufacturing resource virtualization include resources involved in the manufacturing system such as manufacturing hardware devices, networks, software, application systems, etc., as shown in Fig. 2.4.

In a cloud manufacturing system, users face a virtualized manufacturing environment. This approach reduces the degree of coupling between users and resources. Through virtualization technology, a manufacturing resource can form multiple isolated "virtual devices." Multiple manufacturing resources

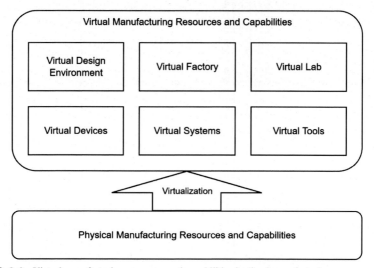

FIG. 2.4 Virtual manufacturing resources and capabilities in cloud manufacturing.

can also be combined to form a larger "virtual device." Virtualization enables real-time migration and dynamic scheduling of manufacturing resources when needed. Virtualization technology enables simplified representation and access of manufacturing resources and capabilities and unified management. Resource virtualization is the basis for realizing the servitization and collaboration of manufacturing resources.

2.3.4 Service-orientation

Cloud manufacturing systems bring together large-scale manufacturing resources and capabilities. With resource virtualization, physical resources become virtual resources. Then through service-oriented technology, virtual resources can be encapsulated as manufacturing services. Manufacturing services can be combined with each other to form various services required in the manufacturing process, such as design services, simulation services, production and processing services, management services, and integration services. The purpose of servitization of manufacturing resources is to provide users with high-quality, cheap and on-demand manufacturing services. On-demand services are mainly reflected in two aspects. First, the centralized use of decentralized resources can be realized through the on-demand aggregation service of cloud resources. Secondly, the decentralized use of centralized resources is realized through the on-demand split service of cloud resources. The manufacturing mode based on the services of manufacturing resources and their combinations has the characteristics of standardization, loose coupling, and transparency. These features can improve the openness, interoperability, agility and integration of manufacturing systems.

The characteristics of cloud manufacturing services include: on-demand dynamic architecture (providing manufacturing services anytime and anywhere according to user needs), interoperability (supporting interoperability between manufacturing resources and manufacturing capabilities), collaboration (manufacturing-oriented multi-user collaboration, large-scale Collaboration of complex manufacturing tasks), heterogeneous integration (supports the distribution of heterogeneous manufacturing resources, and the integration of capabilities), super, fast, and unlimited capabilities (can quickly and flexibly form various services to respond to needs), full life cycle smart manufacturing (serving the whole life cycle of manufacturing, using intelligent information manufacturing technology to achieve full-process intelligent manufacturing across stages).

Cloud manufacturing can provide manufacturing enterprises with "more, faster, better and more economical" services on demand anytime, anywhere, and supports manufacturing enterprises to add "products" and "services" as the leading "integration, synergy, agility, green, service-oriented, intelligent" new economic growth mode development. Cloud manufacturing can support the realization of various advanced manufacturing modes (such as agile manufacturing, concurrent engineering, virtual prototype engineering, mass customization, lean manufacturing, etc.) and improve the market competitiveness of enterprises.

2.3.5 Collaboration

Collaboration is a typical feature of advanced manufacturing mode, especially for the manufacture of complex products. The cloud manufacturing system enables manufacturing resources and capabilities to form flexible, interconnected, and interoperable "manufacturing resources as a service" modules through information technologies such as standardization, normalization, virtualization, servitization, and distributed high-performance computing. Through collaborative technology, these cloud service modules can dynamically realize the whole system, the whole life cycle, all-round interconnection, interoperability, and interoperability to meet the needs of users.

In addition to synergy at the technical level, cloud manufacturing also provides comprehensive support for the dynamic collaborative management of agile virtual enterprise organizations, enabling multi-agent on-demand dynamic construction of virtual enterprise organizations, as well as organic integration and seamless integration in virtual enterprise business collaboration.

2.3.6 Intelligence

Another typical feature of the cloud manufacturing system is to realize the whole system, the whole life cycle, and all-round in-depth intelligence. Knowledge and intelligent science and technology are the core supporting the operation of the cloud manufacturing service system. Manufacturing cloud not only brings together various manufacturing resources and capabilities, but also brings together various knowledge and builds a cross-domain multidisciplinary knowledge base. As the manufacturing cloud continues to evolve, so does the scale of knowledge accumulated in the cloud. Knowledge and intelligent science and technology penetrate into all links and levels of the entire manufacturing life cycle to provide intelligent support.

Under the cloud manufacturing model, knowledge and intelligent science and technology provide support for the "full life cycle" of two dimensions, as shown in Fig. 2.5. One is to manufacture full life cycle activities. The second is the full life cycle of manufacturing resource services. On the one hand, knowledge and intelligent science and technology have penetrated into all aspects of demonstration, design, production and processing, experiment, simulation, operation and management in the whole life cycle activities of manufacturing. This can provide the kind of cross-disciplinary, multidisciplinary and multi-specialty knowledge required. On the other hand, knowledge and intelligent science and technology are integrated into all aspects of the full life cycle of manufacturing resource services, including: resource/capability description, release, matching, combination, transaction, execution,

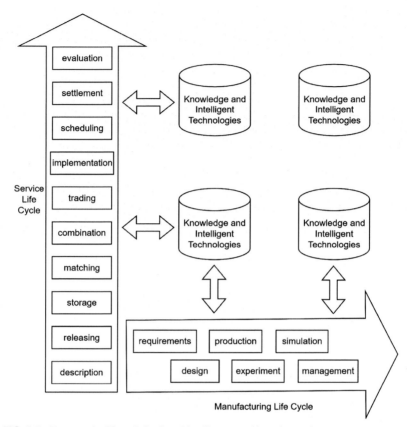

FIG. 2.5   Framework of knowledge-based intelligent manufacturing modes.

scheduling, settlement, evaluation, etc. Knowledge and intelligent science and technology cover each coordinate point in the plane of these two dimensions, providing all-round intelligent support for cloud manufacturing.

2.4   Conclusion

The cloud manufacturing has been widely accepted as a promising mode of future manufacturing. This chapter gives a brief introduction to concept and main features of cloud manufacturing. With the increasing popularity of 3D printing, the combination of cloud manufacturing and 3D printing will be a popular business model in the future, which leads to a customized on-demand production mode with the low cost characteristics of mass production. All services in the cloud manufacturing platform, such as cloud design, cloud simulation, cloud scheduling and cloud logistic, can be used to support 3D printing. In the next chapter, we will discuss the realization of the integration of 3D printing and cloud manufacturing.

References

[1] B.H. Li, L. Zhang, S.L. Wang, et al., Cloud manufacturing: a new service-oriented networked manufacturing model, Comput. Integr. Manuf. Syst. (in Chinese) 16 (1) (2010) 1–8.

[2] L. Zhanglin, Y. Luo, F. Tao, et al., Cloud manufacturing: a new manufacturing paradigm, Enterp. Inf. Syst. 8 (2) (2014) 167–187.

[3] L. Zhang, J.G. Mai, F. Tao, L. Ren, Future manufacturing industry with cloud manufacturing, in: Cloud-Based Design and Manufacturing (CBDM): A Service-Oriented Product Development Paradigm for the 21st Century, Springer, London, 2014.

[4] B.H. Li, L. Zhang, Cloud Manufacturing, Tsinghua University Press, July, 2015.

[5] L. Zhang, X. Xu, D.Z. Wu, Cloud manufacturing: the future of manufacturing, the 2018 World manufacturing forum report, Villa Erba, Cernobbio, 2018, pp. 64–65.

Chapter 3

3D printing with cloud manufacturing

3.1 3D printing in the cloud manufacturing environment

3D printing is a digital rapid prototyping manufacturing mode that integrates new material technology, 3D computer graphics design technology, and numerical control processing technology. Different from the traditional Subtractive Manufacturing processing technology (such as cutting with machine tools), 3D printing generates 3D entities by printing and stacking materials layer by layer [1]. Based on the received 3D model files, 3D printing technology realizes full-automatic stacking processing, which has the characteristics of low cost, short process and digitization. It has bright prospects in realizing the networked, intelligent and cloud based manufacturing industry. As one of the core enabling technologies of industry 4.0, 3D printing can meet personalized rigid requirements and will effectively promote a mass-customized production for the public.

Historically, 3D printing is an old proposition and is unfading with the new technologies. Early in the 1860s, Franois Willeme published the patent photo sculpture, which is regarded as the 3D printing idea prototype. In the 1980s, several molding process techniques based on rapid prototyping have been exploited, such as SLA (Stereo Lithography Appearance), FDM (Fused Deposition Modeling), and SLS (Selective Laser Sintering). In 1993, Emanual Sachs, an MIT professor, produced some 3D parts by rapid prototyping and named this technology 3D printing. With the continuous breakthrough of new technologies, the performance of 3D printers has been significantly improved for nearly a decade. Today, more additive manufacturing processes (especially new synthetic and self-growth materials, multi-nozzle technology) have emerged on the market. These new manufacturing methods are derived from the above three basic process categories of SLA, FDM, and SLS.

Generally, offline 3D printing follows the processing flow shown in the Fig. 3.1. Firstly, the 3D design model is established through the model design tool, and then the 3D design model is transformed into 3D printing model. There

Customized Production Through 3D Printing in Cloud Manufacturing
https://doi.org/10.1016/B978-0-12-823501-0.00012-2
27

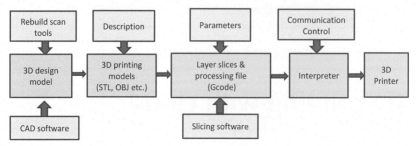

FIG. 3.1 A general production process of 3D printing.

are several kinds of 3D file formats, such as STL, OBJ, AMF and 3MF, model files in different formats can be converted to each other; Then the 3D printing model is divided into section data layer by layer, The section data of each layer includes parameters of the 3D printing actuator, including the motion track, motion speed, feeding speed, heating temperature, etc., all section data form a processing file, which is called slicing. The formats of processing files commonly used for 3D printing include Gcode, SLC and CLI. It is difficult to convert these files between different formats, but they can be obtained by slicing 3D printing model files. Finally, the interpreter on the printing device converts the Gcode file into a control command that can be executed by the 3D printer, to drive the layer by layer printing of the printing device [1,2].

The new generation of information and communication technologies such as the Internet of Things, cloud computing, big data, and new generation communication technology drives manufacturing technology changes with the industry 4.0 era. 3D printing is one of the core technologies driving the development of industry 4.0. It thrives and integrates cutting-edge technologies in multiple disciplines such as information technology, precision machinery, materials science, modeling and simulation. 3D printing is a digital manufacturing technology, its product life cycle is carried by digital information, so it is more suitable for cloud-based production mode.

Cloud manufacturing provides an ideal platform to take full advantages of 3D printing to realize customized mass production as discussed in the previous chapters. Having enough 3D printers on the manufacturing cloud platform will form a new manufacturing mode. This kind of distributed manufacturing mode will be more robust. Even if some printers don't work because of unpredicted uncertainties, others can replace them to provide services still. More importantly, based on the cloud platform, users can produce personalized products at home or office.

At present, there are some commercial platforms of 3D printing, for example, 3D systems [3], 3D hubs [4], Shapeways [5], and Stratasys [6], Mohou 3D [7], Nanjixiong [8], which can provide users with design customization, model correction, resource leasing, and knowledge provision, and other services.

However, the existing commercial 3D printing platforms have two disadvantages. On the one hand, the 3D printing production process of the platforms is not intelligent enough, many activities in the process needs to be completed manually. Judging from the detailed labor division in today's manufacturing industry, even people in the industry can hardly complete the entire manufacturing process of 3D printing based on their personal ability and financial resources. It requires a large team or organization to provide relevant knowledge and equipment as support. On the other hand, since the production process is invisible to and beyond the control of users, the production received often do not meet users' expectations. In particular, the platforms cannot make full use of a large number of potential 3D printing resources to realize low-cost mass customized production in the real sense.

This book aims to introduce a 3D printing service mode supported by a cloud platform based on cloud manufacturing (can be call as cloud 3D printing platform in short) to produce customized products intelligently or automatically and help users to easily participate in the whole life cycle activities of 3D printing.

In the 3D printing service mode, by using servitization integration technology based on cloud manufacturing, functions, models and devices in Fig. 3.1 will be encapsulated into services, and all steps in the printing process will be connected automatically in series. Compared with existing commercial platforms, this cloud 3D printing platform has many new functions and tools such as a 3D model generation, 3D printer access and online monitoring, 3D printing service evaluation and management, and on-demand service matching and scheduling. In short, this platform is committed to effectively integrates the entire production process of 3D printing products, as shown in Fig. 3.2. The benefits of 3D printing with cloud manufacturing include but not limited to:

- Quick response to the individual user's personalized needs;
- Customized design and group innovation;
- Customized production with low cost;
- High flexible production;
- Knowledge and Big Data Support.

3.2 Production with cloud 3D printing platform

3.2.1 The architecture of the cloud 3D printing platform

To sufficiently embody the philosophy of customized production based on 3D printing service in cloud manufacturing, the cloud 3D printing platform should have the features as follows:

(1) Rent online with hardware: Users can not only submit 3D model files to get customized products using offline service, but directly operate the

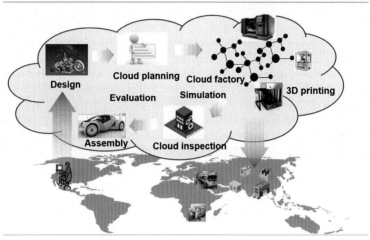

FIG. 3.2 The production process of 3D printing products.

equipment online and monitor the running status as same as a local device operation.

(2) Optimal service on-demand use: Users can search and choose appropriate 3D printing services according to various requirements, such as service type, machining precision, machining speed, material, queue status, price, etc.

(3) Production management in the cloud: Users and Service providers can arrange production planning and manage machines in the cloud, combining with other manufacturing information management tools, e.g., ERP, MRP, etc.

(4) Distributed cloud services cooperation: When a task of mass customization is submitted to the cloud, thousands of manufacturing companies or individual service providers will promptly respond to the needs. And then, it can be finished with distributed 3D printers in the cloud.

A cloud 3D printing platform can be established based on the concept model of cloud manufacturing introduced in Chapter 2. Fig. 3.3 is the architecture of the cloud 3D printing platform to support customized design and production [9].

The cloud 3D printing platform is divided into five layers: (1) The lowest layer is 3D printers and adapters that will be connected to the platform; (2) The second layer from the bottom is the assess adaption layer, to collect operation information of devices, support device monitoring, security management. (3) The third layer provides service encapsulation and management to support service description, publishing, and registration in the platform. (4) The fourth layer includes various tools, such as model design, model evaluation, model matching and composition, scheduling and process optimization, simulation, pricing, etc. (5) The fifth layer is the application layer, and users can

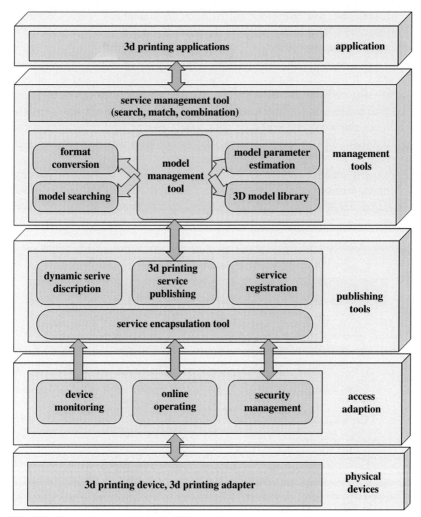

FIG. 3.3 The architecture of the cloud 3D printing platform.

publish their requirements. Application scenarios can also have a library of application cases.

(1) Physical devices layer

All kinds of 3D printing devices will connect to the platform through an adapter. From the perspective of platform manufacturing resources, the physical device layer's resources include various 3D printing devices and the available 3D printing adapter with a universal external access function. Currently, most mainstream 3D printers such as stereolithography appearance (SLA), fused deposition modeling (FDM), 3D printing

(3DP), selective laser sintering (SLS) on the market have been accessed to this cloud 3D printing platform through an adapter.

(2) Access adaption layer

Access adaption layer realizes efficient passing of printing resources' data. It includes two aspects. One is the adapter layer, and the other is the service layer. As shown in Fig. 3.4, The adapter layer can receive, analyze and execute control instructions transmitted by the service layer. This layer can also send real-time data, video, photos, and other information through a data collection system. The service layer uses web services technology to encapsulate various management functions such as service request management, user privilege management, accounting management, service queues management. This layer realizes remote procedure call of 3D printing device management services through service interfaces.

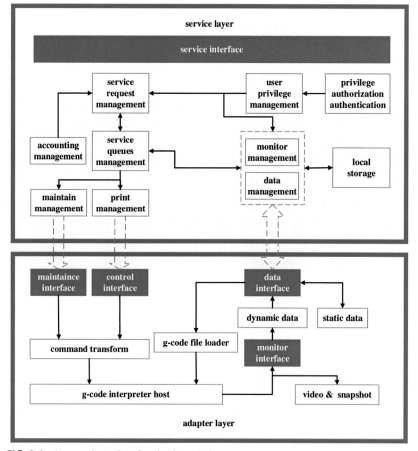

FIG. 3.4   Access adapter interface implementation.

Thus, the cloud manufacturing concept of "centralized use of decentralized resources" is realized. We will discuss further details of this layer in Chapter 5.

(3) Publishing tools layer

The publishing tool layer implements the description, registration, and communication of services through the service encapsulation tool's network service technology. The process of service publish is as follows. First, web services definition language (WSDL) realizes dynamic service description through a complete set of service interface description specifications. These service components are registered to the platform service management center through universal description, discovery, and integration (UDDI) technology. Finally, the service's functional description is loaded into the service management center's index in the form of a label. In this way, different service components from developers' can be managed and scheduled uniformly.

(4) Management tools layer

The management tools layer provides data standard management tools for the application layer to support applications invocation of models and services. This layer includes the model management tool and the service management tool.

The model data in the model management mainly comes from two sources: (1) The model generated by designers and users through online design tools; (2) Standardized models provided by resource providers. These models are standard STL, OBJ, AMF, 3MF, and other format files. For the management of model information, model management tools need to have model storage, evaluation, transformation, and query functions. As a hub, service management tools can connect the application layer with the publishing tool layer. Service management tool needs to have the ability to search, match, and compose various services.

(5) Application layer

The main idea of application layer design is to run through the whole life cycle production process of 3D printing products in servitization. Through servitization, users of the platform can directly participate in each process of 3D printing production. On this layer, 3D printing service applications have been developed to provide support for activities in the 3D printing life cycle, such as 3D model design, printing process monitor, supply-demand matching and task scheduling, 3D printing credibility evaluation, 3D printing process management, security and privacy services.

3.2.2   The functional structure of the cloud 3D printing platform

For clearly expressing the information interaction and collaborative process between modules in the whole platform, the information interaction between modules is described in Fig. 3.5.

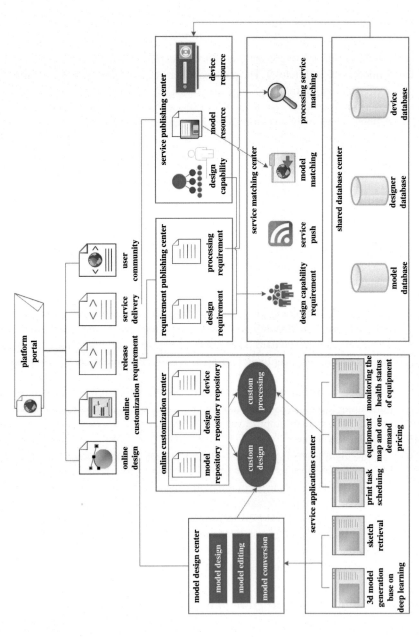

FIG. 3.5 Functional structure chart of the cloud 3D printing platform.

The cloud 3D printing platform database center includes three types of databases: model database, designer database, and device database. The model database is used to store various 3D model information. The designer database is used to store the qualification and ability information of 3D printing model designers. The device database stores specification and parameter information of the 3D printing device. These design capabilities and manufacturing resources are encapsulated in a service-oriented form and published to a service publishing center. When the user posts a printing task requirement, the matching module of the service matching center will match the design resources according to the design requirement. Users can view, modify or design models with the help of designers or application tools. Then the matching module will find most suitable printers and finish the printing task through a set of customized service applications.

3.2.3 Standard for 3D printing platform

A 3D printing task in a cloud platform may require several service providers' contributions to achieve the desired outcome. These workflows may need to adapt to the requirements specific to that outcome swiftly. To do so, a flexible and transparent interface structure is required. Without interface standards, information exchanges between customers and additive manufacturing service providers and between collaborating additive manufacturing service providers often require ad hoc and expensive manual intervention. Inconsistent descriptions of the characteristics of the services provided can also create confusion, misunderstanding, and rework.

The Framework for an Additive Manufacturing Service Platform (AMSP) was published by ISO [10]. It identifies interfaces and their key characteristics where standards can contribute to formalizing the interfaces for submission, design, and creation of 3D printing products. It also regulates the 3D printing service provider's manufacturing ability and effectively promotes collaboration between customers and service providers in a cloud platform. A Framework defining a general functional architecture based on the identified requirements is proposed. The AMSP standard can be used for reference in the construction of cloud 3D printing platforms.

3.3 Advantages of cloud 3D printing services

3.3.1 High customization

Although mass customization provides a way to meet the different needs of as many users as possible through requirements refinement and modular reorganization, the modular combination method has granularity limitations after all. Moreover, in the mass customization mode, users often accept and choose products passively, which means that the limited personalized needs of users

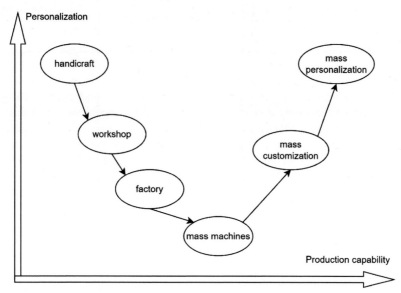

FIG. 3.6   The development of mass customization of manufacturing.

are met. Fig. 3.6 shows the development of mass personalization of manufacturing. In the cloud manufacturing environment, personalized customization based on 3D printing is not only a modular combination, but also the demand for personalized matching of design resources and services in the cloud platform. By designing personalized product models and matching corresponding 3D printing equipment and required materials, the cloud manufacturing platform can customize products according to processing requirements, so as to meet the highly personalized customization needs of users.

3.3.2   High agility

In the traditional mass production process of products, product design often needs to take into account the constraints of processing technology and processing flow. For the simulation of part design and assembly, it is also necessary to process hand sheet samples to correct the design, and finally finalize the 3D model and processing drawings. In the production stage, it is necessary to design the mold, try out the mold, and repair the mold. After the mold is finalized and the sample processing test is qualified, the mass production of the product can be carried out. In contrast, 3D printing technology eliminates the constraints of traditional process requirements at the design stage. The 3D design model is the product, which is beneficial to improve the design efficiency. In the production stage, the mold trial production process is omitted. The 3D printing equipment automatically superimposes the products layer by layer according to the design model, which reduces the number of links and helps to shorten the processing cycle. With the breakthrough of materials that can be processed by 3D printing technology, more products and objects can be produced by 3D printing. Moreover, the cloud manufacturing model supports the availability of 3D printing services and the availability of printed products everywhere. Therefore, the personalized customized production based on 3D printing in the cloud manufacturing environment can quickly respond to the user's product needs and achieve highly agile manufacturing.

3.3.3 High flexibility

In the cloud environment, the customized production of products can be completed by temporarily forming a service chain in the cloud. Product design can be quickly realized mainly through crowdsourcing and collaborative design of design services. For product processing, cloud workshops and cloud factories can be built in the cloud platform through numerous cloud 3D printing devices. In addition, the cloud manufacturing platform can also build various cloud services such as cloud inspection, assembly, logistics, and maintenance. In the production process, when some services have problems, the faulty services can be replaced by other services through intelligent service scheduling on the cloud platform. Especially in the 3D printing processing stage, many online 3D printing equipment can build a highly flexible processing service portfolio.

3.3.4 High socialization

With the gradual development of production methods from handicraft industry to large-scale machine production, the production scale of products has also expanded from scattered small workshops to large-scale socialized production. In the process of this change, the production cost of the product is also reduced.

The cloud manufacturing platform aggregates numerous manufacturing resources and services to form a huge resource service pool. At the same time, the cloud manufacturing model lowers the threshold for users to participate in personalized customization, allowing many individual users and enterprise users with creativity, ideas, technology and knowledge to have the opportunity to use cloud resources for social production. The creativity of an individual user is limited, but the users of the whole society gather together to have a huge group innovation ability. The processing capacity of a 3D printing device is limited, but the aggregation of many devices can form a huge processing capacity. Therefore, cloud manufacturing technology can aggregate decentralized model resources, 3D printing equipment, designers and other manufacturing resources and services to build a huge-scale socialized production environment.

3.3.5   Low cost

The cloud manufacturing model can reduce the cost of personalization. On the one hand, cloud manufacturing platforms can solve the scarcity problem of high-end manufacturing resources by aggregating scattered 3D printing resources. Cloud 3D printing platforms can provide global sharing and on-demand use of high-end manufacturing resources. This model reduces the cost of purchasing and maintaining high-end expensive resources for personal users. On the other hand, cloud 3D printing platforms bring together a wide range of social resources. The characteristics of socialized resources are large in quantity but low in price. Therefore, the cloud 3D printing platform provides a basis for realizing low-cost personalized customization. Based on the on-demand matching and optimal configuration of resources, the cloud platform adopts transparent and open transaction and billing methods, which have the potential ability to solve the additional costs caused by information barriers and opaque prices in traditional production, and reduces the total cost of manufacturing.

3.4   Conclusion

3D printing is an old proposition and rejuvenates by the new technologies, which is characterized by high efficiency, high customization and low cost compared with the traditional manufacturing mode. As one of the core technologies facilitating industry 4.0, 3D printing will effectively promote a mass-customized production for the public. Meanwhile, 3D printing is a typical digitalized manufacturing technology, so it is more suitable for cloud-based design and production. This chapter describes the idea of customized production mode through 3D printing in cloud manufacturing, gives an architecture of the cloud 3D printing platform. In the following chapters, we will elaborate on technologies related to the functions and applications of the platform.

References

[1] J. Horvath, A brief history of 3D printing, in: Mastering 3D Printing, Apress, 2014, pp. 3–10.
[2] E. Bassoli, A. Gatto, L. Iuliano, et al., 3D printing technique applied to rapid casting, Rapid Prototyp. J. 13 (3) (2007) 148–155.
[3] https://www.3dsystems.com.
[4] http://hubs.com.
[5] https://www.shapeways.com.
[6] https://www.stratasys.com.
[7] http://www.mohou.com.
[8] https://www.nanjixiong.com.
[9] J.G. Mai, L. Zhang, F. Tao, et al., Customized production based on distributed 3D printing services in cloud manufacturing, Int. J. Adv. Manuf. Technol. 84 (1–4) (2016) 71–83.
[10] 3D Printing and Scanning, Framework for Additive Manufacturing Service Platform (AMSP). https://www.iso.org/obp/ui/#ISO:STD:ISO-IEC:23510:DIS:ED-1:V1:EN.

Chapter 4

Model design of 3D printing

Model design is the key step in customized 3D printing. Providing strong support for users' rapid customized design is an important task of a cloud 3D printing platform. Most users of 3D printing platform do not have professional design experience, so it is difficult for them to directly give a professional design model. What frequently happens, a user has an innovative idea but can only give a sketch to describe his/her printing requirement, then he/she expects the platform to help him/her to complete a professional design and print a satisfied product. Even for professional designers, if the platform can provide favorable services to help them design or improve their models, it will be very attractive for them to become a frequent user. This chapter will introduce several functions in the platform related to cusmized model design, they are model management, sketch retrieval and image based 3D model generation.

4.1 3D printing model management

As described in Chapter 3, there are various types of models in different formats involved in the production process of a 3D printing product. A 3D model usually goes through a series of de-dimensional processes from 3D to 1D, as shown in Fig. 4.1. Firstly, designers use computer-aided design software such as Solidworks, Pro/E, UG, etc., to build a CAD model (design model). Secondly, designers use meshing techniques to de-dimensionally discrete the CAD model, a series of simple geometry that divides the model's surface into triangles or polygons. In this step, the CAD model is converted to file formats, such as STL or OBJ. Invented in 1988 by Charles Hull, founder of 3D Systems, the STL file format is now the industry standard for interface file formats for CAD/CAM systems worldwide. It is the most common 3D printing model file format supported by 3D printers. The OBJ file is a standard 3D printing model file format developed by Alias Wavefront for its workstation-based 3D modeling and animation software "Advanced Visualizer." Currently, almost all mainstream 3D printers support these two file formats. Thirdly, these grid files use slicing software to describe each layer of graphics information after slicing. In this period, meshing the polygonal graphic closure test in the previous stage is essential because it is impossible for an unclose single-layer graphic to

Customized Production Through 3D Printing in Cloud Manufacturing
https://doi.org/10.1016/B978-0-12-823501-0.00009-2

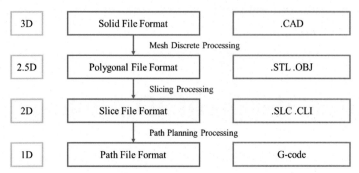

FIG. 4.1 A general process of traditional 3D printing geometrical model.

complete path planning. Therefore, many related studies are trying to solve and optimize this problem of the polygonal graphic closure test. Finally, the 2D slice file will be converted into a 1D path motion instruction or grating instruction for the 3D printer to complete the mechanical printing action.

For CAD models, different software gives different model formats. To solve the problem of data exchange of different CAD models in different formats, the International Organization (ISO) for Standardization provides a set of standards for data exchange, therein the standard exchange of product data mode (STEP) provides a neutral interoperability mechanism that does not depend on specific modeling software. STEP has been favored by much mainstream CAD software providers, such as CATIA, Pro/E, Solidworks, AutoCAD, UG, etc.

For 3D printing models, STL file format has become the global 3D printing industry standard, and is the most common file format for 3D printing. OBJ format is developed by a set of workstation based 3D modeling and animation software "advanced visualizer." It mainly supports polygon models and is suitable for data exchange between 3D printing models. Most 3D CAD software and 3D printers support OBJ format. However, STL and OBJ cannot adapt to the new 3D printers with multi-color, multi-material, multi-process. Additive manufacturing file format (AMF) [1] and 3D manufacturing format (3MF) [2] were developed for incorporating those multi-dimension information.

Compared with the diversification of models, the number of models increases more rapidly with the progress of 3D printing model design technology and the popularity of 3D printing in recent years. In a cloud 3D printing platform, model is one of the most valuable resources in a 3D printing platform. There are two main sources of 3D printing models in a cloud 3D printing platform: (1) models submitted by users, which are usually designed or prepared by professional users; (2) models provided or collected by the platform, which are used as materials for professional or non-professional uses to customize their models. All the models need to be well managed through a model library, which will be the basis of customized design. In order to efficiently manage 3D models and support customized 3D model design, we designed a cloud based 3D

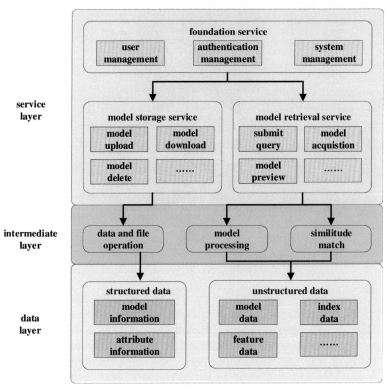

FIG. 4.2 3D model library framework.

printing model management library framework to complete basic management functions, such as uploading, storing, matching, and converting, and realize rapid sketch matching and optimization and modification of 3D design and printing models. The framework is shown in Fig. 4.2, which has three levels: data layer, middle layer, and service layer.

(1) **Data layer**: The data layer is used to store both structured and unstructured data. The structured data is metamodel description information of the 3D model, including model information and attribute information. The model information is descriptive information that will automatically generate when the model is created by users, including model ID, name, storage location, creation time, etc. The attribute information is the model's parameter information for the 3D printer, such as the material type, texture, color, etc. Unstructured data is based on the refined classification of model image content, including model data, feature data, and index data. Among them, Model data is a 3D model source file readable by a 3D printer, including standardized STL, OBJ, AMF, 3MF, and other formats. Feature data is coordinate position data used to describe the model outline drawing.

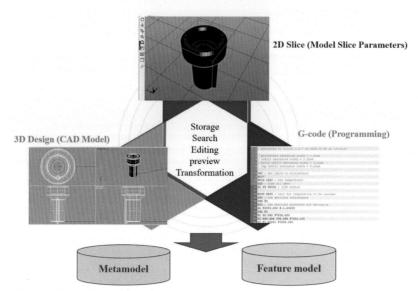

FIG. 4.3 3D printing model management framework.

Index data provides searchable address labels for model data and feature data. In the library, there are different kinds of models with standardized model description formats, including 2D or 3D CAD models, STL, OBJ, AMF or 3MF files, G-code files, etc. These model information are further subdivided and distributed to the metamodel library and feature model library, as shown in Fig. 4.3. The metamodel library is used to store structured data for the data layer. The feature library stores unstructured data.

(2) **Intermediate layer**: The intermediate layer supports model retrieval and model storage of the service layer through the underlying components, such as the module of data and file operation, model processing, and similitude match. The architecture is more scalable by encapsulating and reusing the underlying components.

(3) **Service layer**: In the service layer, various function modules are encapsulated as services through web service technology. A standardized interface facilitates other system's remote procedures call to the layer's functional modules. To ensure model resources security, we have designed foundation services such as user management, authentication management, and system management. On this basis, the model storage service and the model retrieval service will authorize the corresponding service module usage rights to the relevant users according to the user authority manage functions. Model storage services include model upload, model download, and model deletion. The model retrieval service can match the corresponding model resources according to the retrieval requirements to achieve rapid retrieval, including the functions of the model query, model acquisition, model preview, and more.

4.2   Sketch based 3D model retrieval

Customized manufacturing starts from customized design. Whether it can help users to realize their creativity is the key to customized manufacturing. As we stated in the beginning of this chapter, when a user draw a sketch, the platform needs to help the user to complete a professional design that can be used for printing. For this purpose, we developed a sketch- based 3D model retrieval approach based on the model library to help users quickly obtain satisfied 3D models [3]. The approach consists of two components, i.e., 3D printing model aggregation and 3D printing model retrieval. 3D printing model aggregation is a dynamic process of orderly organization and efficient management of models. 3D printing model retrieval is a process of matching models in the library. These two technologies will be discussed below.

1. **3D printing model aggregation**
    (1) **Model aggregation process**: The metamodel library and the feature library shown in Fig. 4.3 are established to realize 3D printing models' aggregation. The 3D printing model aggregation process enables the systematic organization and efficient management of models to realize storing, sharing and reuse of models. The process of model aggregation is shown in Fig. 4.4. First, for a 3D model file submitted by a user to the platform, the structured information of these files are extracted through the normalized description template and the information is stored in the metamodel library. Then the unstructured data of the 3D model is extracted and saved into the feature library after preprocessing. The metamodel library and the feature library compose the data layer of the 3D print model management framework.

    Most of the existing model resources have been stored in the metamodel library and feature library through the model aggregation process. All of them need to be processed by the model aggregation process for the user's new model resources. In the process of model aggregation, the model

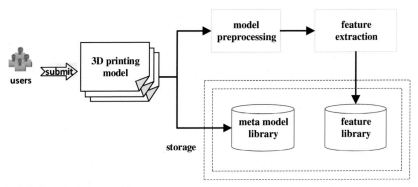

FIG. 4.4   3D printing model aggregation process.

preprocessing and feature extraction process will bring substantial computational costs. Dynamic aggregation mechanism is adopted here. This mechanism performs timing batch processing on these models by setting thresholds. In this way, the model library will be divided into several sub-model libraries according to different description information of these models.

(2) Model preprocessing: Before storing the model in the feature library, the parameters such as the position, size, and direction of the model need to be normalized. Model preprocessing is the process of transforming and standardizing these model parameters. Commonly computer-aided design software, such as AutoCAD, Solid Works, UG, Pro/Engineer, etc. can be used to design 3D models. The file formats suitable for reading by 3D printers derived by CAD software are mainly STL and OBJ.

In the model preprocessing stage, coordinate processing of STL and OBJ files is required. Model's coordinate preprocessing process is divided into the following three steps:

(i) Calculating the geometric center coordinates of the 3D model. It is necessary to establish a unified three-dimensional coordinate system, and the geometric center coordinates of the model are coincident with the origin of the coordinate system by a translation operation.

(ii) Model pose vector calibration. To align all 3D models in a uniform orientation by using principal component analysis (PCA) to do pose vector calibration.

(iii) To make the 3D model have a uniform scale. After translation, rotation, and scaling, all models will have a uniform position, orientation, and size for subsequent view acquisition and feature extraction. The geometric transformation formula of the 3D model preprocessing process is as follows:

$$\widehat{G} = \left\{ g \mid g = S^{-1} MR(g - \overline{h}), g \in G \right\} \tag{4.1}$$

In (4.1), G represents a collection of original vertex coordinates in the 3D model, g is model's original vertex coordinate, S^{-1} is the scaling factor matrix of a model, M is a transposed matrix in the form of a diagonal matrix, R represents the rotation matrix, $\overline{h}$ represent the geometric center coordinates of the model.

(3) Feature extraction: Feature extraction is the process of operating the model after being preprocessed to obtain the unique information that can represent the model. The Features of the 3D model are mainly divided into two categories, one is the feature vector, and the other is the topology map. The feature vector is a set of values that express the geometric features of a model, corresponding to coordinates in three-dimensional space. The topology map can well reflect the space topology structure of the model. The feature extraction technology of 3D printing models can be divided into the following three categories: statistical feature extraction, topology feature extraction, and view feature extraction. A 3D model can be projected into a set of 2D images at different viewpoints. The acquired feature set is integrated by the view feature extraction method, and the extracted feature set can establish a one-to-one mapping relationship with the three-dimensional model. This approach greatly reduces the computational complexity of the retrieval process.

2. 3D printing model retrieval

In the study of 3D printing model retrieval, the keyword-based retrieval technology and the content-based retrieval technology are the mainstream in current studies. However, keyword-based retrieval technology is gradually replaced by content-based retrieval technology due to the large amount of manual work (Note: content-based retrieval technology is a research branch on large-scale digital image content retrieval). In the research of content-based 3D model retrieval technology, according to the different methods of feature extraction in the retrieval process, it is mainly divided into the retrieval technology based on model information and the retrieval technology based on view information. However, the above retrieval methods all require the 3D view of the 3D printed model for matching operations. Retrieving massive 3D models in a cloud manufacturing environment will bring substantial computational costs. Therefore, it is necessary to improve the retrieval model based on the above technologies.

In terms of retrieval performance and retrieval effect, 2D model retrieval is more mature than 3D model retrieval technology. Therefore, the extension of 2D model retrieval to 3D through image synthesis technology is the mainstream research direction in the field of graphics retrieval. In this chapter, we use the view projection technique to represent the features of the 3D model with a group of the 2D image of this model's view. To reflect the characteristics of customized design, we developed a sketch-based 3D model retrieval method [3]. Users are allowed to describe their requirements through hand-drawn sketches. Which is an easy way for unprofessional users to naturally express their ideas. The hand-drawn sketch characteristics can link to the 2D view features of the model. Adds flexibility.

Fig. 4.5 shows the frame of the sketch-based 3D model retrieval approach. The user first draws the sketch shape of the desired model and submits the sketch. The platform will then automatically run the model retrieval process in the background and return several highest matching 3D views models for user selection.

In the model aggregation process, we extract the 3D model view features and store the feature data in the corresponding feature library. When the user draws a sketch, the selected projection viewpoint position is arbitrary, so all potential projection viewpoints should be considered when selecting the

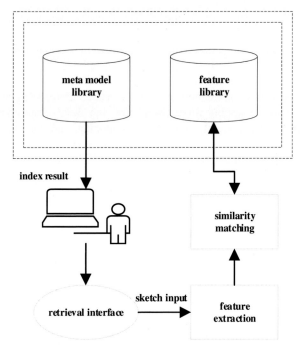

FIG. 4.5 The frame of the sketch-based 3D model retrieval approach.

projection viewpoint of the 3D model. We summarized the location distribution characteristics of potential projection viewpoints by analyzing the user's drawing history data. This method will significantly reduce the computational cost of searching for all potential projection points.

The closed outline is an important feature of the 3D model, which embodies the basic shape of the 3D model. The closed outline feature of the hand-drawn sketch input by the user can well reflect the model's shape information. Therefore, the 3D model feature extraction method based on a model's closed outline features is very suitable for sketch retrieval. Then, the model's closed outline features are drawn according to the projection viewpoint selected by the user and stored in the two-dimensional image view set.

In the cloud manufacturing environment, there are many 3D printing models. The number of view sets generated from these models will be scaled up. The mass of the model view will bring enormous challenges to the retrieval of models. To speed up the retrieval, the study based on the bag-of-features (BoF) model solves the problem of fast matching between sketches and view sets (Note: the BoF model is one of the most widely used technologies in computer vision in recent years). In this project, we use the scale-invariant feature transform (SIFT) clustering method to cluster the local feature vectors of images and form a visual dictionary. The specific retrieval steps are as follows:

- Get SIFT feature vectors for 3D models and sketches.
- Extract image features of SIFT feature vector according to Bag-of-Features model.
- Use the K-Means method to form the visual dictionary.
- Generate the histogram vector to describe the model's 3-dimensional closed outline features.
- Match the histogram vector of the 3D model view set and the sketch to get the optimal matching solution set.

A 128-dimensional histogram vector will be generated based on the above techniques, which can uniquely describe a specified model's 3-dimensional closed outline features.

3. Case study

The following case demonstrates the realization process of model aggregation and model retrieval in a cloud platform. Fig. 4.6 is an example of the model outline features extraction for the model aggregation process. For a three-dimensional view of the 3D printed model, taking the model's geometric center as the projection viewpoint, the model's closed outline features from 20 different angles are obtained. In this way, each 3D model can acquire a set of closed outline feature views. Then, all the SIFT features in the view set are clustered using the K-Means method to form the visual dictionary. Finally, each 3D model will obtain a histogram representing the model's 3-dimensional closed outline features through querying the visual dictionary.

In the model retrieval process, the histogram can also be obtained according to the user's sketch model. The system then matches the histogram to the view feature vector in the view set and returns a similar set of 3D models to the user. Fig. 4.7 shows the sketch-based 3D model retrieval approach's actual retrieval effect, which has been embedded in the cloud manufacturing platform as an application tool. When using the tool, the user first draws the desired model's sketch shape and submits it. The system will then automatically run the model

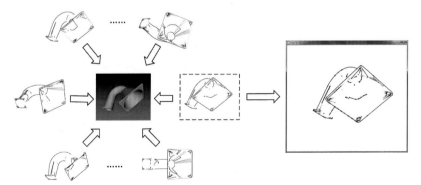

FIG. 4.6 Feature extraction of 3D model's close outline view.

Display sketch

Match result

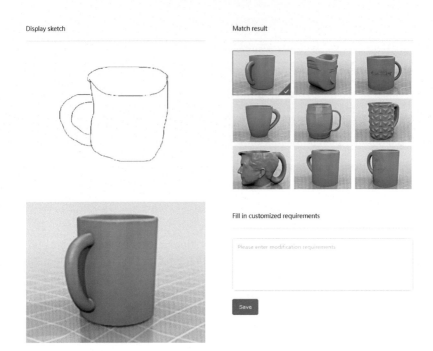

Fill in customized requirements

Please enter modification requirements

Save

FIG. 4.7 The sketch-based 3D model retrieval approach.

retrieval process in the background and return the user with the highest matching eight 3D views models. This sketch-based 3D model retrieval approach achieves the reuse of the model to a certain extent.

4.3 Image based 3D model generation

The design of a 3D model is time and energy consuming and require professional design skills, which leads to a high threshold. In recent years, along with the development of 3D scanning technologies, reverse reconstruction technology based on entity models can dramatically simplify the design process. Several graphics software implements this function, such as insight3D, 3DFLOW, 123D Catch, etc. However, these software's' operating threshold is usually very high and needs to cooperate with a related 3D scanning device to realize the physical model's reconstruction. Therefore, they are not quite general for common users. As the correlative research on the 3D model in Deep Learning has been more and more popular, it is an alternative method to generate 3D shapes with generative models in the deep neural network. For this reason, we carried out research on 3D model generation methods [4] based on big data because we have previously accumulated a large number of 3D models in the cloud manufacturing platform in the study of sketch retrieval approach.

The research on 3D model generation is inspired by the deep representation of 3D model voxels. It can considerably reduce the computing burden. In this way, we only need to use the image of the product to reconstruct the 3D model structure of the entity product. Based on Conditional Generative Adversarial Network (GAN), many models will be used as training samples and input as conditional vectors in GAN. Thus it will have the ability to generate more 3D voxels. Depending on cloud computing clusters' efficient computing power, based on a deep learning algorithm of GAN, these products' images will be automatically transformed into a standard 3D printing model file. In the following sections, we will give the details of the 3D model generation process, experiment and evaluation results.

4.3.1   3D model generation process

The whole 3D model generation process is showed in Fig. 4.8. We leveraged the neural network's learning ability to generate 3D shapes represented in the 3D voxel grid containing simple occupancy information. The input is a single image while the output shape is 32^3 occupancy grids in our network. And finally, the output shape will translate the standardized 3D design file such as STL or OBJ.

To make it easier to annotate image information. The image encoder is donated as E, Generator is denoted as G, the discriminator is denoted as D. We pass the single image into a pre-trained model as a feature extractor to obtain a 2048-dimensional latent vector Z. We feed Z into the volumetric convolutional networks, which serves as the Generator in Conditional WGAN-GP. The image feature serves as conditional input without noise input in our case. The generator tends to learn a map from images to 3D shapes with the supervision of 3D shapes. The generated model Y is further fed into a conditional discriminator. In particular, a single image paired with its corresponding 3D voxel model Y, which is regarded as "real" example, while the image paired with its corresponding output 3D voxel shape from the generator is regarded as "fake" example. We aimed at enhancing the discriminator to learn the matching relationship between image space and 3D shape space. The generator and discriminator are learned in an adversarial way. The discriminator as a critic is learned to distinguish the input pair is real or fake, with respective scores output. And the difference score between real example and fake example represented the distance between real data distribution and generated data distribution. The discriminator output signal is used to update both the generator and discriminator asynchronously. The training process is illustrated in Fig. 4.9.

FIG. 4.8   3D model generation process.

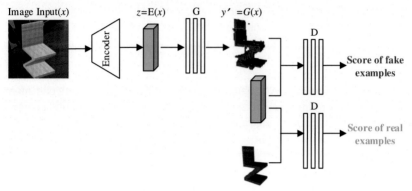

FIG. 4.9 Training a Conditional GAN to map images to voxel models.

(1) Image encoder

The pre-trained deep learning models are called image feature extractors, such as VGG-Net, Res-Net, Google-Net, etc. These pre-trained models with convolutional architecture are usually used in classification tasks where images must be classified into one of 1000 different categories. Each of the whole pre-trained models can be divided into feature extracting layers and classification layers. The information we need is the output of feature encoder layers. In this case, we chose ResNet-50 to map a single image to feature vector. On the one hand, this model's total parameter number is smaller than VGG-19's. On the other hand, the output feature map is $1 \times 1 \times 2048$, which can be easily reshaped to a feature vector passed to the following generator and discriminator.

(2) Generator

We design a 3D convolutional neural network to generate 3D objects inspired by 3D-GAN. Distinct from traditional Conditional GANs, we omit the noise vector input because we found that the generator simply learned to ignore the noise. And we desire the generator to produce deterministic outputs. The image feature vector is computed via a Fully-Connected layer in the generator, output with an embedding vector of the same size. Therefore, the generator could adjust the external signal via the FC layer parameters updating. Then we reshape the embedding vector into $2 \times 2 \times 2 \times 256$ as the source feature map of the 3D deconvolutional network. As illustrated in Fig. 4.10, the network includes five fully convolutional layers with

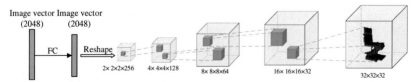

FIG. 4.10 Generator architecture.

kernel size $4 \times 4 \times 4$ and strides 2. The generator maps a 2048 dimensional vector extracted from a single image via a pre-trained encoder to a 32^3 cube, representing a 3D shape in voxel space.

(3) Discriminator

The discriminator is arranged to classify whether the estimated 3D shapes are plausible or not by scoring the 3D voxel models. As stated above, the conditional discriminator's input signal includes not only the 3D shape only but also the image feature vector. The shape of input 3D voxel cube is 32^3 whereas the image feature is represented as a vector. The difficulty is how to combine these two signals in distinct dimensionality in the discriminator. We reference the method from an application called generative adversarial text to image synthesis. We process the input 3D voxel data with three 3D convolutional layers of kernel size $4 \times 4 \times 4$ and strides 2, but each followed with leaky ReLU. We reduce the image feature vector's dimensionality through a fully connected layer to a 200 dimensional embedding vector. When the spatial dimension of 3D convolutional layers' output is $4 \times 4 \times 4$, we replicate the embedding vector spatially and generate a matrix in the size of $4 \times 4 \times 4 \times 200$. Then process the combined data via a 3D convolutional layer with the same configuration as above. Finally, we get a final score from the discriminator. The detailed architecture as shown in Fig. 4.11.

(4) Object reconstruction loss

We aim to train a Conditional WGAN-GP [5] to map a picture to a 3D shape. Besides the objective function L_{gan} for Conditional GAN, an object reconstruction loss L_{recon} is introduced.

Modifications of the binary cross-entropy loss function are used for the generator. The original binary cross-entropy weights false positives and false negatives equally. Nevertheless, most of the voxel space tends to be empty, and the network can drop into a local optimum point by outputting most negatives. In this set, we inflict a higher penalty on false-positive than false-negative results by assigning a hyper parameter τ which weights

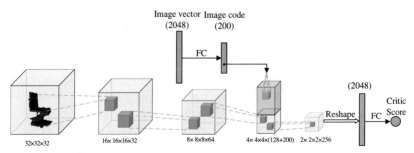

FIG. 4.11 Discriminator architecture.

the relative importance of false positives against false negatives as shown in the following formula (4.2):

$$L_{recon} = -\tau y \log (y') - (1 - \tau)(1 - y) \log (1 - y') \tag{4.2}$$

Where y is the target value in $\{0, 1\}$. y' is the output value for each voxel from the generator.

(5) GAN loss

In the Conditional WGAN-GP settings, the loss function is slightly modified, and detailed definitions can refer to WGAN-GP:

$$L_{gan}^g = -\mathbb{E}[D(y'|z)] \tag{4.3}$$

$$L_{gan}^d = \mathbb{E}[D(y'|z)] - \mathbb{E}[D(y|z)] + \lambda\mathbb{E}\left[\left(\left\|\nabla_{\hat{y}}D(\hat{y}|x)\right\|_2 - 1\right)^2\right] \tag{4.4}$$

Where $\hat{y} = \varepsilon x + (1 - \varepsilon)y'$, $\varepsilon \sim U[0, 1]$, lambda controls the trade-off between the original WGAN loss and the gradient penalty. As stated in the original paper, the advised value of lambda was 10. z represents the input image feature vector, while y' is the generated value in $(0, 1)$ for each voxel from the generator, and y is the target value in $\{0, 1\}$.

There are two loss functions, L_{recon} and L_{gan}^g to synergistically optimize the generator in our case. Minimizing L_{recon} tends to learn the overall 3D shapes while minimizing L_{gan}^g tends to decrease the high-frequency noise of predicted output. To jointly optimize the generator and guarantee the two losses in a scalar, we assign the weight η_1 to L_{gan}^g and η_2 to L_{recon}. Overall, the loss function for the generator is showed as follows:

$$L_g = \eta_1 L_{gan}^g + \eta_2 L_{recon} \tag{4.5}$$

During training, we set η_1 to 0.5, and η_2 to 100.

(6) Training

The whole architecture is trained end-to-end in a supervised way. The image encoder provides a feature vector for both the generator and discriminator. They are not needed to fine-tune in the training process as the parameters are updated asynchronously. To be specific, if we update image encoder parameters in a feed-forward manner, the image encoder will drop into an unstable state and influence the generator and discriminator's learning process. The image feature extracting ability of pre-trained ResNet-50 is convincing, and we adapt the feature vector through the fully-connected layer in the generator and discriminator.

The discriminator has more computing power than the generator. There is no need to worry about the gradient vanishing problem. Therefore, we update the discriminator five times while updating the generator one time. We set the learning rate of both G and D to 0.001, and use a batch size of 128, we use ADAM optimizer, with $\alpha = 0.0001$, $\beta_1 = 0.5$, $\beta_2 = 0.9$. The training procedure as follow:

Algorithm 4.1 3D-ICGAN, our proposed algorithm. We use default values of $\lambda=10$, $n_{critic}=5$, $\alpha=0.0001$, $\beta_1=0.5$, $\beta_2=0.9$, $\tau=0.85$, $\eta_1=0.5$, $\eta_2=100$.

Require: The gradient penalty coefficient λ, the number of discriminator iterations per generator iteration n_{critic}, the batch size m, Adam hyper-parameters α, β_1, β_2, the reconstruction loss hyper-parameter τ, the loss for generator update hyper-parameters η_1, η_2.

Require: ω_0, initial discriminator parameters. θ_0, initial generator parameters.

1: **while** θ has not converged **do**

2:    **for** $t=1, \ldots, n_{critic}$ **do**

3:      **for** $i=1, \ldots, m$ **do**

4:        Extract image feature vector z, a random number $\varepsilon \sim U[0,1]$.

5:        $y' \leftarrow G_\theta(z)$

6:        $\hat{y} \leftarrow \varepsilon y + (1-\varepsilon)y'$

7:        $L_D^{(i)} \leftarrow D_\omega(y') - D_\omega(y) + \lambda\left(\left\|\nabla_{\hat{y}}D(\hat{y}|z)\right\|_2 - 1\right)^2$

8:      **end for**

9:      $\omega \leftarrow \text{Adam}\left(\nabla_\omega \frac{1}{m}\sum_{i=1}^{m} L_D^{(i)}, \omega, \alpha, \beta_1, \beta_2\right)$

10:   **end for**

11:   Extract image feature vector $\{z^{(i)}\}_{i=1}^{m}$

12:   $y' \leftarrow G_\theta(z)$

13:   $L_{recon}^{(i)} \leftarrow -\tau y \log(y') - (1-\tau)(1-y) \log(1-y')$

14:   $L_G^{(i)} \leftarrow -D_\omega(G_\theta(z))$

15:   $\theta \leftarrow \text{Adam}\left(\nabla_\omega \frac{1}{m}\sum_{i=1}^{m} \left(\eta_1 L_G^{(i)} + \eta_2 L_{recon}^{(i)}\right), \theta, \alpha, \beta_1, \beta_2\right)$

16: **end while**

---

(7) Data generation

For training tasks, we need to obtain a large amount of training data. We collect 3D shapes from the Shapenet database, which covers 55 common object categories with about 51,300 unique 3D models. Besides, we render all the models into 15 viewers, at elevation of 30 degrees and 15 azimuth angles from 0 to 360 degrees, in increments of 24 degrees. The original image data has no background. We will render the images on a colorful background selected from public datasets such as the SUN database. We converted the 3D shape into 32^3 voxel grid using software such as binvox.

In conclusion, the 3D printing model generation process based on Generative Adversarial Network can be described in Fig. 4.12.

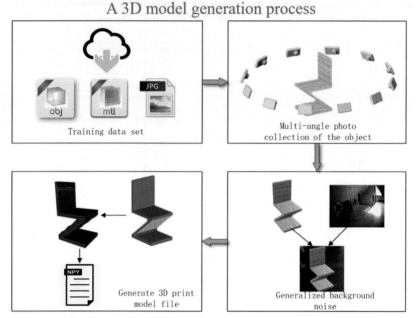

FIG. 4.12   A 3D printing model generation process based on Generative adversarial network algorithm.

4.3.2   Experiment and evaluation

This section will evaluate the 3D Conditional WGAN-GP and introduces the procedures to access a printable file of 3D shape from voxel grid output of deep generation model.

The 3D shape is represented in a voxel grid while generated by 3D Conditional GAN in a voxel grid, but the printable 3D shape is polygon mesh format.

We aim to obtain physical 3D shapes rather than a group of visual 3D voxels that do not describe their coordinates [6]. It is necessary to convert the representation of voxel to polygon mesh represented by the coordinates of their vertices [7]. Several uncontrolled outliers away from the main voxel grid structure generated from our deep learning model appear at times. These redundant voxels will negatively impact the efficiency of 3D printing as needless. However, we don't have to worry about it because these negative effects will be repaired in the slicing process. We presented a connected-domain-detection method that employed connectivity of 26 in our 3D binary grid and implemented this method in MATLAB. In this way, we eliminated the outlier voxels treated as noises of the 3D voxel model (see Fig. 4.13).

The generated 3D voxel models are denoised via extracting the largest connected component of the generated object. However, it is insufficient to generate a printable 3D object to convert voxel data to a polygon mesh [8]. As we know, the file format most often used in 3D printing is STL, which describes only the surface geometry of a three-dimensional object. In this case, we apply

Input Image

Voxel
model

Denoised
model

Polygon mesh

FIG. 4.13   Eliminate the outlier voxels of the 3D voxel model.

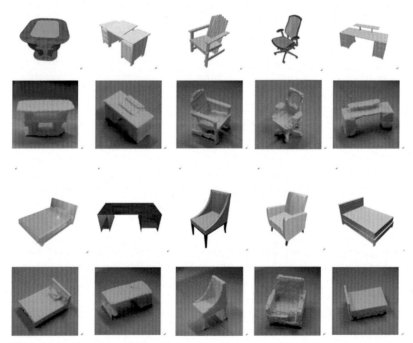

FIG. 4.14   Input images and printed objects.

the Lewiner marching cubes algorithm to find surfaces in 3D volumetric data. So the STL file can be exported for 3D printing.

The generated object is printed, and is illustrated in Fig. 4.14. The model possesses the ability to generate a 3D printable object as long as images are submitted.

This approach can significantly reduce the designer's workload, although the designer still needs to reconstruct and repair the generated model. The user only needs to submit the physical photos of the product to be printed. Through the powerful computing power provided by the cloud platform, the photos will be output to the designer as STL files, thereby reducing the product manufacturing cycle.

4.4 Conclusion

To realize customized 3D printing based on the cloud platform, 3D model design technologies are discussed in this chapter. A model library framework is introduced to manage a large number of 3D models with various formats, which will be a foundation of developing customized design technologies. And the model library itself has great value. Based on the model library, two approach that can help achieve customized design are introduced and two application approach and corresponding tools are developed accordingly, they are sketch based 3D model retrieval and image based 3D model generation. With the support of the two tools, users can get desired printable 3D models by drawing sketch's or submitting images.

These model design approaches give two ways to customized 3D model design. Nevertheless, they are far from perfect. There are still lot of work need to be done in the future. For example, for the 3D model retrieval tool, image feature extraction and matching will consume a lot of computing resources. How to use the Hadoop distributed file system (HDFS) for model distributed storage and use MapReduce parallel computing framework to build the model feature library, which is one of the issues worthy of follow-up research. Besides, some parameters adjustment and optimization methods in the image retrieval algorithm need to be strengthened to improve the model's search efficiency and matching accuracy. For image based 3D model generation approach also needs to be continuously improved, for example, by the lights of transfer learning theory, the voxel model's noises can be better eliminated by using a large number of product models' voxel features as the training set.

In practice, the application effects of both the tools depends on the number of models in the model library to a large extent. The construction of 3D printing model library for different industries is an important work in the future.

The model design technology also relies on the progress of 3D printing hardware equipment. With the continuous emergence of new generation of 3D printers such as multi-nozzle, high-precision material, the model description framework of 3D models will be further updated to meet the production requirements of new printers. Accordingly, the model management and design technologies will be updated following the changes of 3D printers.

References

[1] ISO, Specification for Additive Manufacturing File Format (AMF) Version 1.2. ISO/ASTM 52915:2020. https://www.iso.org/standard/74640.html.

[2] 3MF, Consortium Specification. https://3mf.io/specification.

[3] X. Luo, F. Pan, L. Zhang, et al., Study of 3D printing model aggregation and retrieval mechanism in cloud manufacturing, in: Proceedings of IEEE 17th International Conference on Industrial Informatics, Helsinki-Espoo, Finland, 2019.

[4] Z.M. Li, L. Zhang, Y.Q. Sun, L. Ren, Image-based 3D shape generation used for 3D PRINTING, in: Proceedings of Asian Simulation Conference 2018. Beijing, China, Springer, 2018, pp. 539–551.

[5] I. Gulrajani, F. Ahmed, M. Arjovsky, V. Dumoulin, A.C. Courville, Improved training of wasserstein GANs, in: Proceedings of the 31st International Conference on Neural Information Processing Systems, 2017, pp. 5769–5779.

[6] G.K.L. Tam, R.W.H. Lau, Deformable model retrieval based on topological and geometric signatures, IEEE Trans. Vis. Comput. Graph. 13 (3) (2007) 470–482.

[7] P. Daras, A. Axenopoulos, A 3D shape retrieval framework supporting multimodal queries, Int. J. Comput. Vis. 89 (2010) 229–247.

[8] M. Eitz, R. Richter, T. Boubekeur, Sketch-based shape retrieval, ACM Trans. Graph. 31 (4) (2012) 1–10.

Chapter 5

3D printing resource access

5.1 Classification of 3D printers

According to Wohlers Report 2020, researchers note that the global additive manufacturing sector has been grown at an annual average rate of 23.3% over the 4 years from 2016 to 2019 [1]. The total value of the industry was 11.86 billion dollars in 2019. Currently, 3D printing, as terminal products, accounts for the largest proportion at 30.9%, as shown in Fig. 5.1. This shows that 3D printing has achieved a qualitative change from rapid prototyping to rapid manufacturing.

To achieve the manufacturing platform's transformation, access to manufacturing resources into the cloud service platform is the most important thing. In 3D printing, there is a high correlation between the device classes and the associated materials that can be manufactured by these machines. As we all know, the technical execution of 3D printing processes is carried out by means of layer build processes. However, many different technological processes generate a solid layer that forms parts by attaching and connecting adjacent layers [2]. These technological processes have gradually evolved into process families, and they are only suitable for one or some unique materials.

With the development of 3D printing technology, many different types of 3D printing equipment have been developed. According to different usage scenarios, 3D printers can be divided into industrial, quasi-industrial, and desktop 3D printers. Since 1987, more than 100 machines for direct digital production have been developed and introduced to the market that follow the principle of layer manufacturing using the basic physical principles. Out of these machines, the most often used have been selected and listed below according to the main families of printing materials and molding techniques of additive manufacturing.

(1) Polymerization
- (Laser) Stereolithography (SL)
- Polymer Printing/Jetting

(2) Sintering/Melting
- (Selective) Laser Sintering ((S)LS)
- Selective Laser Melting (SLM)
- Selective Mask Sintering (SMS)
- Electron Beam Melting (EBM)

Customized Production Through 3D Printing in Cloud Manufacturing
https://doi.org/10.1016/B978-0-12-823501-0.00001-8

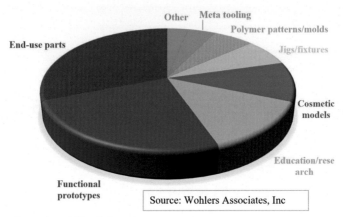

FIG. 5.1 Proportion of 3D printing usage.

(3) Layer Laminate Manufacturing (LLM)
- Paper Lamination
- Plastic Lamination

(4) Extrusion Fused Layer Modeling (FLM)
- Fused Deposition Modeling (FDM)

(5) Powder Binder 3D Printing
- Three-Dimensional Printing (3DP)

(6) Aerosol Printing

This list contains the generic names and commonly-used abbreviations. Based on these families, manufacturers have developed different processes and machines. Current 3D printing materials mainly include liquid resin, plastic filament, polymer material, plaster, metal and so on. The liquid resin is instantly heated and formed by laser irradiation, and is used to process fine model samples, handicrafts, etc. Its characteristics are that the processed products have high precision and moderate prices. Plastic wire can be used to process model samples, toys, handicrafts, plastic shells, etc. Its features include ease of use, low cost, and are widely used in desktop 3D printers. The polymer material is extremely thermoformed by a laser. Due to its high strength, it can be used to process samples with certain strength requirements, such as prototype parts and product parts. Plaster is available in liquid form and powder form. Due to its environmental friendliness and ease of use, it can be used in medical applications such as orthopedics or dental prosthetics. Metals are often used in the manufacture of products in the aerospace industry due to their high strength. The titanium alloy is a high-end industrial metal material with excellent performance. Although titanium alloy is half as dense as steel, they are far stronger than most other alloys. China's titanium alloy laser forming technology has developed rapidly, and titanium alloy materials have been used in the manufacture of large-scale main bearing components in China.

However, the technical bottleneck for the further application of 3D printing technology is still limited by the types and properties of materials. Firstly, the variety is limited, especially the advanced materials suitable for daily life are not rich enough. Personalized customization of daily necessities has unlimited market potential. In addition to the relatively mature materials such as metal powder, gypsum, polymer materials, liquid resin and plastic filaments, various materials such as food, textiles, and ceramics need to be developed, and even theoretically. That said, almost all materials can be used for 3D printing. Secondly, the safety of materials to the human body needs to be guaranteed. The 3D printing process of most materials is a process of heating accumulation or laser sintering. Whether the processed products can meet the requirements of safe use still lacks unified evaluation and inspection standards. Thirdly, the strength properties of materials need to be improved. In industrial applications, most products need to meet certain impact resistance and durability index requirements, so it is required that the materials used to manufacture products and molded products must have certain strength requirements.

In the future, with the further development of 3D printing technology and the breakthrough of new materials, 3D printing technology will be promoted to enter daily life, education, medical treatment, manufacturing, and other industries on a larger scale. The number of new devices is increasing steadily on the market. Based on the new generation of 3D printers, materials can be mixed during the printing process, and thus different materials within one part can be generated [3]. Besides, common material types like plastics, metals, and ceramic materials, also medical or biological materials can be processed into products. As mentioned above, there are more and more types of 3D printers on the market. If we want to access them into the 3D printing platform, we need to classify them from the perspective of the device access category. There are two main types of methods to access 3D printers through cloud platforms: accessing 3D printers based on adapters and accessing 3D printers based on sensors, which we would discuss in detail as follows.

5.2   Accessing 3D printers based on adapters

The real-time data of 3D printers can be efficiently transmitted to the cloud platform by using adapters. The framework of the online adaptation system is shown in Fig. 5.2. In order to connect the 3D printing equipment to the cloud platform and provide online services, three key functional modules are needed including the adapter access module on the adapter side, the service management module on the platform side, and the communication module between the platform side and the adapter side. We discuss these three technical methods in this section respectively. Finally, we will discuss the distributed slicing based on adapters.

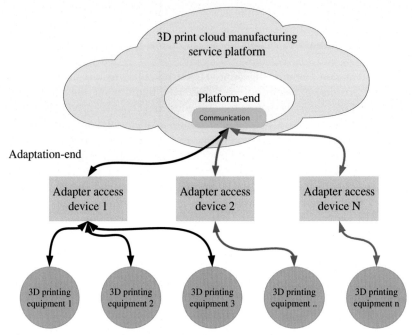

FIG. 5.2 The framework of the online adaptation system.

5.2.1 Adapter access module on the adapter side

Fig. 5.3 shows the system framework of the adapter access module which includes equipment real-time data collection, task execution progress data collection, user permission management, communication between adapters and 3D printing devices, online operations of 3D printers. The adaptation access module is based on the embedded hardware and Linux operation system to run various functional modules. Multiple 3D printing devices are able to be connected to the adapter through multiple USB interfaces. And then the communication between 3D printers and the cloud platform is implemented through the network.

The interface unit and program module of the 3D printing adaptation access module mainly includes a data processing unit, a network interface unit, a video interface unit, and a device interface unit, as shown in Fig. 5.3. The data processing unit runs the 3D printing service programs, and its main functions include storing static data of 3D printing equipment (such as accuracy, model, size, etc.), collecting dynamic data of 3D printing equipment (such as temperature, vibration, image, processing progress, etc.), receiving processing files sent from the cloud platform to the 3D printing equipment, executing, suspending, stopping, adjusting printing tasks through the cloud platform, permission management, and equipment security management.

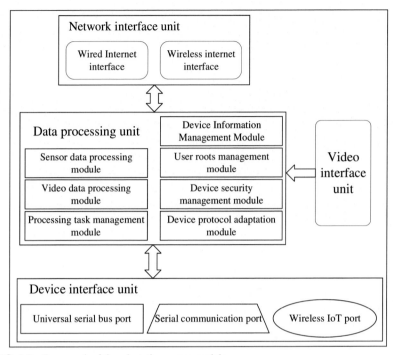

FIG. 5.3 Framework of the adaptation access module.

The network interface unit is used to connect the data processing unit with the Internet, including a wired Ethernet interface and a wireless network interface. The video interface unit is used for real-time video monitoring of the running status of the 3D printing equipment. The device interface unit is used to connect several 3D printing devices, including multiple USB ports, multiple serial ports, and wireless data transmission interfaces, as shown in Fig. 5.4.

FIG. 5.4 Physical connection between the adaptation access module and 3D printers.

The data processing unit is to run multiple program management modules on the embedded system, including sensor data processing module, video data processing module, equipment information management module, processing task management module, equipment security management module, user rights management module, and equipment protocol adaptation module, etc.

Among these modules, the sensor data processing module is to read sensor data such as temperature, current, and voltage from the 3D printing devices through the data cable, and transmit the data to the cloud manufacturing platform server. The function of the video data processing module is to obtain real-time video images through a video camera and transmit the video data to the cloud manufacturing platform server. The function of the processing task management module is to execute instructions from the cloud manufacturing platform such as receiving and reading tasks to form a task queue, adjusting the task queue, and executing, pausing, and stopping tasks. The function of the device information management module is to record and store the basic information of the 3D printing devices connected to the adapter. The basic information of 3D printers includes printing accuracy, printing materials, model size, etc. The function of the user permission management module is to analyze and process the remote operation permission of users to 3D printers. The function of the equipment safety management module is to ensure the safety of equipment operation through real-time data analysis. The function of the device protocol adaptation module is to automatically select the corresponding communication protocol for data communication with the device according to the difference of the device model, receive and process the real-time online remote operations on 3D printing devices by the cloud manufacturing platform server. These real-time remote operations include connecting and disconnecting printing nozzles, three-dimensional movement, nozzle temperature setting of the heating platform. The 3D printing adaptation access device is used to realize: (1) Uploading the real-time sensor data and video data of the 3D printing equipment to the cloud manufacturing platform; (2) Managing and using 3D printing equipment online through the cloud manufacturing platform; (3) Ensuring the security and fault protection of the operation authority of 3D printing equipment in the cloud platform; and (4) Connecting multiple different types of 3D printing devices.

5.2.2 Service management module on the platform side

Fig. 5.5 shows the system framework of the platform side in the adapter-based 3D printer accessing system, including registration management of multiple adaptation access devices, cloud-based management of static and dynamic data of 3D printing equipment, printing task queue management, and remote-control interface.

1. Accessing registration and management

The accessing registration/management module includes request token module, access token module and device scan module. The request token

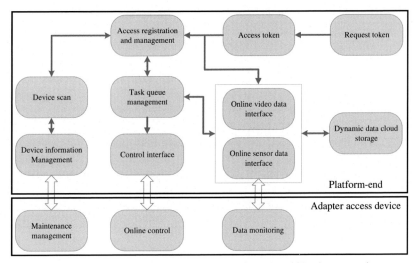

FIG. 5.5   System framework of the platform side in the adapter-based 3D printer accessing system.

module is used to control the issuance of temporary tokens, and to establish a series of encrypted data of trusted registration information for the adaptation access device. Before issuing the temporary token, the identification code of the adapted access device needs to be verified first. After the verification is passed, a temporary token is generated and issued to the adaptive access device. The adaptation access device writes the token into the platform database to complete the registration. The access token module is used to control the issuance of access tokens. The access token is encrypted data used to ensure secure data communication between the adaptive access device and the cloud platform. Before the access token is issued, the identification code and authorization token of the adapted access device need to be verified. After the verification is passed, an access token is generated and issued to the device. The device scan module is used to process the port scan result sent by the device to the cloud platform. Port scan results are encoded in JSON format. After receiving the scan results, the platform preprocesses the content and uses the JSON module in the runtime to parse the data. The content obtained by the analysis mainly includes the serial port name and serial port hardware serial number connecting the 3D printer and the adapter access device. This serial number is an important parameter to distinguish multiple 3D printing devices which were connected to the same adapter access device.

2. **Remote monitoring and equipment operation module**

   The remote monitoring and equipment operation module is responsible for sending an instruction to the adaptation access device to turn on the camera, which is used to request a snapshot of the working conditions of the 3D printing equipment, and display it after receiving the snapshot. At the same time, the service sends a request to transmit video stream data, and displays the real-time video after receiving the stream data. In addition, the remote monitoring and equipment operation module also issues operation requests for each actuator of the 3D printing equipment, such as movement, feeding, material withdrawal, nozzle heating, and constant temperature plate heating. After the request is passed, the user will be granted the operation authority to execute the control instructions from the user.

3. **Dynamic data cloud storage module**

The dynamic data cloud storage module is responsible for storing the real-time data of 3D printing equipment in the cloud, including unformatted video streaming data, photos, and formatted real-time data of running operations. The formatted data is uniformly encoded in JSON format, and is stored in a distributed Couchbase database that is easy to expand and highly reliable.

4. **Task queue management module**

The task queue management module includes access device queue, user queue, queue creation, task information, task failure processing, task creation, task cancellation, task downloading, and completed task processing.

(1) The access device queue module is responsible for acquiring all the queue information of the specified access device, including the name and serial number of each queue, the name and serial number of the device it belongs to, and so on.

(2) The user queue module is responsible for obtaining the queue information of the specified user, including all queue names, serial numbers, and serial numbers of the devices they belong to. The queue creation module is responsible for generating the task queue of the 3D printing device according to the specified name, and setting the operation authority level of the queue, including closed level, open level, and audit level. The closed level indicates that the owner of the own device has the operation authority. Open level means that no user restrictions are set. The audit level indicates that the user who has been audited and confirmed has permission to operate the queue.

(3) The queue creation module is responsible for generating the task queue of the 3D printing device according to the specified name, and setting the operation authority level of the queue, including closed level, open level, and audit level. User restrictions, the audit level is that the user who has been audited and confirmed has the permission to operate the queue.

(4) The task information module is used to obtain the task information of the 3D printing equipment queue. The task information includes task serial number, printer serial number, task name, queue serial number to which the task belongs, task source file path, task status, task progress, task expected processing time, task actual processing time, etc. This interface is used to analyze the historical data of processing tasks

of 3D printing equipment. In addition, this interface can obtain the information of the specified task of the specified printing device. The task sequence number is included in the client request. The adaptation access device matches the required task information according to the task sequence number. The format of the returned data is the same as the "All Task Info" request. In order to ensure the validity of the information, before returning the data retrieval result, the adaptive access device compares the information with the data backed up in the cloud and performs qualification verification.

(5) The task failure processing module is responsible for processing when a print task error is found. When the 3D printing equipment performs the processing task, there is a possibility that the printing task may fail due to internal errors or forced offline. In this case, the adaptation access device sends the fault processing information to the cloud platform. The platform side finds the corresponding task information and resets the task, and at the same time sends a reset instruction to the 3D printing device, so as to ensure that the task status information of the platform side and the adapted access device is synchronized. This feature is used to improve the robustness of the system, and to solve the problem of unreliable communication that may be caused by errors in printing tasks. The create task module is responsible for generating tasks according to the specified information. Transfer information such as task files and queues required for processing tasks to the 3D printing equipment. After receiving the creation request, the adaptation access device performs corresponding processing according to the user level and the authority of the adaptation access device. If the user level satisfies the operation authority, the task will be automatically added to the task queue for processing.

(6) The task cancellation module sends a task cancellation instruction to the adaptation access device, requesting to cancel one or more tasks of the specified 3D printing device.

(7) The download task module processes the task information that has been downloaded to the 3D printing device by the adapted access device. Downloading refers to transferring the processing file to the 3D printing device. The process of downloading the task file is: (1) The adaptation access device receives the file URI and user operation level sent by the cloud platform. When the operation authority is satisfied, the device applies to the cloud platform to transfer the file according to the sent file URI, and also includes token information for authentication; (2) After the file transfer is successful, the adapted access device transfers the file according to the corresponding letter protocol. Download to the 3D printing device; (3) The adaptation access device sends the successfully downloaded task serial number to the platform. The platform side updates the task status and 3D printer status accordingly, and records the download completion time. This interface process ensures reliable operation and status synchronization of task downloads between the adaptation access device, the 3D printing device, and the cloud platform.

(8) The completion task processing module is responsible for receiving and processing the completion task information sent by the adaptation access device. First, the platform side updates the status of the corresponding task in the cloud platform according to the task completion instruction sent by the device. At the same time, the platform sends a notification to the user that the printing task has been completed, prompting the user to take away the completed printed parts. Then, when the pickup is completed, the adaptive access device automatically recognizes that the pickup is completed according to the sensor information, and sends a pickup completion instruction to the platform and the 3D printing device. Finally, the platform side analyzes whether there are other tasks in the queue of this device. If there are no other tasks, the status of the printing device is set to "idle," otherwise, the next task download instruction is sent to the 3D printing device.

5. Operation data management module

The task progress information is updated based on the received data such as task progress, nozzle temperature, and hot bed temperature periodically sent by the adaptive access device. The data is encoded in JSON format. The platform side parses the data content and displays it, and at the same time saves the obtained data to the cloud platform database. This function is used to support dynamic data monitoring of 3D printing equipment.

6. Device Information Management

(1) Reading the user's 3D printing device list: Retrieve the 3D printing device information list associated with the specified user and the corresponding serial number of the adapted access device on the platform.

(2) Reading the detailed information of the printing device: Retrieve the detailed information of the 3D printing device with the specified device serial number on the platform including device resource number URI, device name, device running status, device type, hardware port name, port serial number, communication baud rate, etc.

(3) Updating device information: Users can edit and modify the basic information of the device through the cloud platform. The platform side updates the modified information synchronously to the adaptation access device.

5.2.3 Communication between platform and adapters

From the perspective of data flow, the adaptive access device is the data intermediary between the 3D printing device and the cloud platform. The adaptive access device is responsible for efficient, reliable, and secure two-way data exchange between the 3D printing device and the cloud platform. In the 3D printing cloud service platform, the processing service demander has two basic needs: one is the online processing service call of the equipment, and the other is the real-time monitoring and control of the operating data. The application of the former is characterized by the need for loose decoupling between the platform and the device. In this case, it is necessary to ensure the security of the device during concurrent use. The latter application is characterized by the need to ensure real-time and reliable. Based on the above application characteristics, the producer-consumer model can be used for online processing task invocation of equipment. A hybrid mode of polling and push can be used for real-time data monitoring and control.

1. **Communication mechanisms for online printing tasks**

The communication mechanism based on the producer-consumer pattern is shown in Fig. 5.6. When the platform side communicates with the adapter side, the platform side is the producer, and the adapter side is the consumer. When the adapter communicates with the device, the adapter is the producer, and the device is the consumer. Its basic operating mechanism includes the following three aspects:

(1) Loose coupling of platforms and devices

In the traditional application of 3D printing equipment, a user often can only arrange printing tasks for one or several devices. Therefore, users need to consider the idle situation of the device to make task arrangements. And

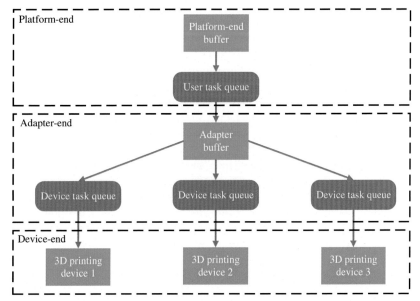

FIG. 5.6   Communication mechanism based on producer-consumer pattern.

during the printing process, the user needs to pay close attention to the execution situation, such as whether the printing fails, whether the task is completed, and so on. This pattern is a tightly coupled approach. In the cloud platform, the adapter is used to analyze and judge the running status of the connected device, and transmit the running status of the device to the cloud platform. The cloud platform makes resource allocation based on the availability of the adapter. In this way, the user does not need to pay attention to the situation of a certain device, but only needs to submit his own task arrangement. At this time, the task queue of the device is allocated by the adapter.

(2) Concurrency support

The "concurrent" here means that multiple adaptive access devices execute the printing task of one user. "Synchronization" is the arrangement of the order in which tasks are performed. The solution is to build a task file buffer that adapts to the access device on the platform side. The adapter side builds the task file buffer of each device. In this way, the communication intervention of the platform can be reduced in the operation of the device.

(3) Security mechanism

In order to improve the communication efficiency between the adapter and the device, a non-authenticated communication method is used between the two. The resulting problem is a lack of security for communications. In order to improve the security of communication, the following measures are adopted: (1) When the adapter terminal communicates with the platform terminal, the message contains token information; (2) The adaptor performs operation permission processing on the operation request of the cloud user.

2. Communication mechanism for real-time data monitoring and control

Fig. 5.7 shows the communication mechanism for real-time data monitoring and control based on the data push mode. Communication mechanisms for real-time data monitoring and control fall into two categories: polling and push. Polling is a data request sent by the adapter to the device at a specific time interval (such as every 1 s), and the request needs to be confirmed by a response, and then the device returns the latest data to the adapter. This method is characterized by simplicity, reliability, and suitability for active port scanning. Therefore, this method is often used for communication between the adapter and the 3D printing device. However, if it is applied to the platform side and the adapter side, this method will have obvious disadvantages. The platform side needs to continuously send requests to the adapter side. However, the authentication information in the request consumes bandwidth resources. The bandwidth occupation caused by a large number of invalid information at the adapter end is particularly obvious. Another communication mechanism is push. In the push mode, the platform side and the adapter side only need to do a handshake. After that, a fast and reliable channel is formed between the two. The adapter can directly send data to the platform at any time. The platform side can also directly send control commands to the adapter side.

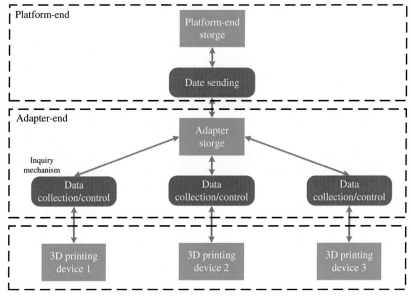

FIG. 5.7 Communication mechanism based on push mode.

5.2.4 Distributed slicing task execution based on adapters

Slicing requires a lot of computation, which can easily overload the slicing server, and cannot guarantee the quality and response speed of each user's slicing tasks. The traditional way is to use the slice server on the platform side to undertake all slice computing tasks. If the slicing calculation is concentrated on the slicing server on the platform side, the slicing calculation itself will take up a lot of CPU and memory when facing a large number of users and 3D printing devices in the platform. Tested on a computer with i5, 8G memory, the execution time of the algorithm is generally in the range of a few seconds or more than 10 s according to the parameter settings during slicing. In the cloud platform, if the user needs to complete the 3D printing task, he must pass the slicing step. When a user actually submits and executes a machining task, most of them will repeatedly construct their own machining task. In addition, a machining job may contain a large number of solid models. The cloud platform has a large number of users with such needs. In order to ensure the execution of slices within a certain response time, huge computing pressure is imposed on the cloud platform slice server. As the number of users in the cloud platform increases exponentially, servers with higher computing power are required to ensure the effective completion of user slice computing tasks.

We propose a distributed slicing method suitable for multi-process 3D printing equipment in a cloud platform. By defining the adapter interface file, the problem faced when online slicing is solved. Massive equipment needs to deal with a variety of processes. The slicing algorithm should be able to adapt to different specific types of 3D printing equipment with various processes, and ensure that the slicing algorithm can be expanded after connecting to new equipment. Using the slice layer interface file as the interface between the two solves the problem of adaptation between a variety of 3D printing devices and slice software connected to the cloud platform. In addition, when a large number of users access at the same time, the task of the slice server is heavy. The proposed scheme should be able to realize the scalability of the slicing algorithm service and ensure that the slicing task can be efficiently performed when the number of users increases. At the same time, the solution can ensure that the processing files can be correctly parsed and executed on the optimally matched 3D printing equipment, which meets the individual needs of users for different slicing algorithms, strategies, support structure design and other algorithms. After a large number of 3D printing devices are connected to the platform, the entire life cycle of 3D printing services can be realized. The adapter plays the role of data transmission in the cloud platform. The data transfer only occupies the IO interface of the adapter, and its computing power is idle most of the time. Therefore, we can consider improving the utilization of idle computing resources of the adapter, so as to share the computing pressure of the slice server.

For different types of 3D printing equipment, the format of the processing file generated during the slicing process has various forms. First of all, different types of 3D printing equipment are affected and restricted by the process, and the formats of the processing files used are different, such as the CLI format with the slice level as the processing track and the G code in the line segment scanning format. In addition, the 3D printing equipment of the same process type generates different processing files due to its control circuit firmware, which cannot be universal. For example, the processing files of 3D printing equipment of FDM process type have Gcode, S3g, x3g and other variants. For a specific Gcode, the meanings of its G and M codes corresponding to specific devices may also be different. Its specific control commands and related characters are closely related to the type of specific equipment.

In order to generate a processing file usable by a corresponding 3D printing device, some specific parameters in the slicing process need to be set by the owner of the printing device according to the specific printing device. Distributing the slicing process to specific equipment makes the whole process more natural and makes the logic simpler. In this way, the computing power of the adapter can be effectively used, and the parallel execution of slice calculations can be realized for many model slices at the same time, which improves the response speed and simplifies the business process of the cloud platform to complete the 3D printing process. Considering the actual usage scenario, the URL address of the adapter may not be fixed, so a star network topology is adopted. We use the Netty-based remote procedure call method to achieve asynchronous two-way direct communication between each adapter and the central dispatch server. We use multiple task queues to achieve asynchronous execution of tasks. The adapter implements indirect communication through the platform and better realizes the execution of asynchronous tasks.

The communication between the adapter and the platform is mainly realized by the RPC method based on Netty. Netty is an open-source JAVA framework that can be used to rapidly develop reliable high-performance web servers. Using Netty can implement asynchronous, non-blocking, event-driven network programs, which is different from the traditional blocking IO model. The non-blocking IO model does not need to block and wait for the transmission of data packets in the process of data exchange between the sender and the receiver. With its multiplexer idea, it can handle a large number of concurrent connection requests with high performance. Therefore, Netty is chosen to handle the data interaction between the adapter and the platform. The communication protocol is implemented based on TCP. Netty completes the encoding of business objects to binary data and the decoding process of binary objects to business objects. The communication mechanism based on Netty is shown in Fig. 5.8.

A remote procedure call (RPC) can implement the function of accepting and sending commands when calling a remote method locally, just like calling a local method. Through the RPC method and the dynamic proxy mechanism of JAVA, the dynamic proxy of the remote method interface is called locally. The specific method of implementing the interface remotely can realize the communication between the two parties, and is easy to program the business logic of the platform side. In order to obtain the real-time information of the adapter, the platform adopts the heartbeat mechanism to realize the communication between the adapter and the platform. The default heartbeat period is 5 s. The multicast method based on the process type is adopted instead of the broadcast method to reduce the pressure of communication. The communication protocol includes two parts: the protocol header and the data. The communication method of short connection and heartbeat is adopted. Heartbeat messages are actively sent by the adapter at regular intervals. After receiving the heartbeat message, the platform side performs data exchange and returns a response

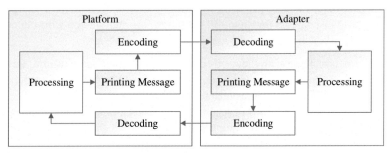

FIG. 5.8   Communication mechanism based on Netty.

message to the adapter. In order to ensure the asynchronous execution of response and message processing, the processing method of multiple message buffer queues is adopted.

The adapter configures the physical property information of the 3D printing device, that is, the service information of the 3D printing device in the platform. After the adapter is powered on and the attribute information of the connected 3D printing device is configured, it will actively send a registration message to the platform through the heartbeat mechanism. The registration message carries the attribute information of the specific 3D printing device. After receiving the registration information of the adapter, the platform stores the device information in the live device cache queue and assigns a globally unique device ID to it. The device ID is returned as a reply message to the registration message. After the adapter receives the reply message, it decodes the obtained global device ID and writes it to the configuration file. This device ID is carried in the following messages such as task request and survival confirmation. After the printer is registered, it will take the initiative to bid on the platform, hoping to obtain tasks that can be performed. In order to ensure the correct execution of slicing tasks in the platform, each adapter and the platform need to cooperate closely, including scheduling each adapter. There are three traditional scheduling strategies as follows.

(1) Centralized scheduling Centrally controls all adapters through the platform, and collects adapter information. Distribute tasks for each adapter by analyzing information about the adapter. This scheduling method has a simple structure and is easy to establish. The disadvantage is that the platform is overloaded, and the collected adapter information is lagging behind, and the real-time performance is poor.

(2) Completely distributed scheduling relies entirely on the adapter to negotiate autonomously to execute tasks. The adapter collects all the information of the task set by itself and combines the analysis of its own information and the communication with other adapters to make a decision. The advantage is that the real-time decision-making is good. The disadvantage is that it is prone to conflict between adapters. And due to the complex algorithm, the load of each adapter is large.

(3) Hybrid scheduling combines centralized scheduling and fully distributed scheduling. The adapters work together by sharing information. The platform is responsible for regularly obtaining information and working conditions of each adapter, and assigning tasks and resources.

To sum up, combined with the needs of the 3D printing cloud platform, a centralized scheduling method can be considered to facilitate the centralized management of information such as adapters, 3D printing models, and user needs. At the same time, the centralized scheduling method can ensure the reliability of printing task execution and equipment security. The platform-side service queue and scheduler mainly include live device cache, slice task cache queue, print task cache queue, slice task agent and slice task scheduler. When the adapter executes the slicing task, it mainly completes four steps: downloading STL, slicing calculation, returning evaluation data, and uploading the interface file. The adapter side mainly includes slice executor and slice task queue. Fig. 5.9 shows the adapter-based slicing task workflow.

5.3   Accessing 3D printers based on sensors

For 3D printing devices that cannot be connected with an adapter or extra information is needed for monitoring the running status or printing process of a printer, we can obtain the parameters of the 3D printer by placing sensors on 3D printers. The system framework of accessing 3D printers based on sensors is shown in Fig. 5.10. The state of the 3D printer is described by the parameters of the sensor, including a temperature and humidity sensor, a vibration sensor [4], an acceleration sensor [5] and so on. The output of each sensor is integrated into the open-source hardware microcontroller Arduido [6]. At this time, Arduido can be used as a personal computer or Raspberry Pi server peripheral hardware [7]. After finishing connecting and debugging of multiple sensors on the Arduino board, installed Arduino on a 3D printer that can work normally. This allows us to collect multidimensional data from the printing process. During the actual use of 3D printing equipment, the 3D printer data obtained by the sensor is uploaded to the cloud in real time, which supports subsequent 3D printer status analysis, task processing, and real-time equipment monitoring.

5.3.1   Sensor selection and application

In order to use sensors to collect 3D printer data, the selection of sensors is very critical. There are many types of sensors that can be selected. The choice of sensors depends on the task you are going to do with the information obtained from the sensors, as well as the type of the printer. We take the printing failure monitoring as an example, which will be discussed in Chapter 6. Combined with the basic information and status parameters of the 3D printer, a batch of sensors can be selected to collect key working status parameters of the printer. Then, the data based on these sensors can be analyzed and its parameters can be used for model training and validation. We use vibration sensor and acceleration sensor to show the process of accessing a 3D printing. The printer used in this study is a fused deposition (FDM) printer, as shown in Fig. 5.11. Fused Deposition Modeling (FDM) is a technology invented by Scott Crump of Stratasys in the United States in the late 1980s. FDM is another widely used 3D printing technology after SLA and LOM. In 1992, Stratasys launched the world's first 3D printer based on FDM technology, marking the commercial stage of FDM technology.

The working principle of FDM is to heat and melt the filamentous thermoplastic material through the nozzle. There are fine nozzles (generally 0.2–0.6mm in diameter) at the bottom of the nozzle. Under the control of the

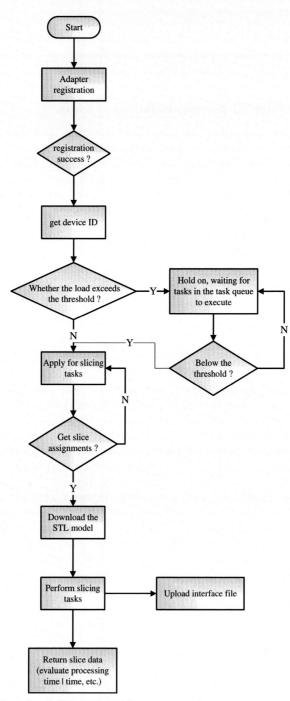

FIG. 5.9   Adapter-based slicing task workflow.

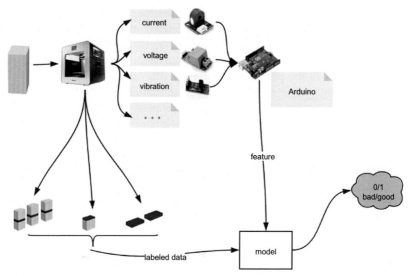

FIG. 5.10 System framework of accessing 3D printers based on sensors.

FIG. 5.11 3D printer based on fused deposition modeling.

computer, the nozzle moves to the designated position according to the data of the 3D model, and extrudes the liquid material in the molten state and finally solidifies. The material is ejected and deposited on the previous layer of cured material. The final product is formed by layer-by-layer accumulation of materials. Before the printer works, data information such as the spacing of each layer of the 3D model and the width of the path must be set. The 3D model is then sliced by the slicing engine (Cura) and the print movement path is generated. Under the control of the computer, the printing nozzle makes the plane movement of the X-axis and the Y-axis according to the horizontal layered data. The vertical movement in the Z-axis direction is completed by the lifting and lowering of the printing platform. At the same time, the material (wire) of the 3D printer is sent to the nozzle by the wire feeding transmission part, and then heated and melted. The material is extruded from the nozzle and bonded to the work surface, which is rapidly cooled and solidified. The material thus printed quickly fuses with the previous layer. As each level is completed, the workbench is lowered by one level. The printer then continues to print the next layer. Repeat this process until the entire object is printed.

It can be known from the working principle of the above FDM that the movement of the nozzle is composed of parts. The X and Y axes move horizontally. And the Z-axis direction does the movement in the vertical direction. The movements in the three directions are driven by stepper motors on the three axes, respectively. According to the working principle of the above FDM, it can be known that the most important working parts to control the movement state of the entire printer are the nozzle part of the 3D printer and the three stepper motors. Therefore, we consider placing sensors for these four components. Fig. 5.12 shows the print head components of the printer. After the printing material is sent here to be melted, it is bonded to the printing platform through the nozzle at the end. Fig. 5.13 shows one of the three-directional stepper motors in the 3D printer, which are responsible for controlling the movement of the printhead in the printing space.

Based on the above analysis, we consider installing vibration sensors and acceleration sensors for the print head and three stepper motors. Among them, since the stepper motors of the X and Y axes are easier to install the sensors, it is considered to install two shock sensors. The Z-axis stepper motor is inside the 3D printer body, so it is necessary to disassemble the body, which is difficult to install. In addition, we place a vibration sensor and an acceleration sensor on the print head part, and the vibration sensor can also detect the vibration signal in the Z-axis direction.

FIG. 5.12 Print head components of the printer.

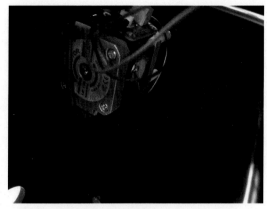

FIG. 5.13 One of the three-directional stepper motors in the 3D printer.

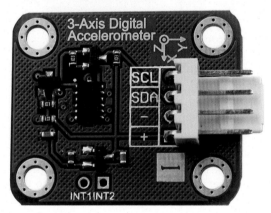

FIG. 5.14 Arduino ADXL345.

Fig. 5.14 shows the accelerometer, which is an Arduino ADXL345. The Arduino ADXL345 three-axis acceleration sensor uses the ADXL345 chip. This chip uses MEMS technology and has SPI and IIC digital output functions. The chip is small and thin, low power consumption, variable range, and high resolution. Its overall dimensions are only 3 mm × 5 mm × 1 mm. When the typical voltage VS is 2.5 V, the power consumption current is about 25–130 μA.

The ADXL345 is ideal for mobile device applications. It can measure static gravitational acceleration in tilt detection applications. It can also measure dynamic acceleration due to motion or shock. The device offers several special detection functions, activity, and inactivity detection. Motion can be detected by comparing the acceleration on any axis to a user-set threshold. The tap detection function can detect single and double vibrations in any direction. Free fall detection can detect if the device is falling. These functions can be independently mapped to one of the two interrupt output pins.

FIG. 5.15 Analog piezoelectric ceramic vibration sensor.

The vibration sensor is an analog piezoelectric ceramic vibration sensor, as shown in Fig. 5.15. The sensor utilizes piezoelectric ceramics to give an electrical signal the reverse process of vibrating. When the piezoelectric ceramic piece vibrates, an electrical signal is generated, which is used in conjunction with the Arduino sensor shield. The Arduino analog port can sense weak vibration electrical signals, and can realize functions related to vibration. Therefore, we can get the six state parameters of the 3D printer through three vibration sensors and one three-axis acceleration sensor placed on the print head and two stepper motors. The values of these six parameters can be recorded at each sampling time. The subsequent functions are completed based on these data.

5.3.2 Sensor fusion

This topic uses Arduino microcontroller unit [8] for sensor fusion. Arduino is an open-source microcontroller controller based on Atmel AVR microcontroller [9]. Arduino has a Processing/Wiring development environment similar to Java and C. In addition to being easy to use, the price of the Arduino microcontroller controller is also very cheap, and the experimental cost is also relatively low. Therefore, this topic finally chooses Arduino to integrate sensors for data acquisition and transmission. As shown in Fig. 5.16, the Arduino Uno is the latest model in the Arduino USB interface series [10]. This is the most widely used model of the Arduino microcontroller control board, and it is also the model selected for this topic. Its processor core is ATmega328, including 14 digital input/output interfaces, 6 analog inputs, a 16MHz crystal oscillator, a USB interface, a power socket, an ICSP interface and a reset button.

FIG. 5.16   Arduino microcontroller unit.

After selecting the hardware, we use the Arduino integrated development kit and use the Arduino Integrated Development Environment (IDE) for the programming development. The Arduino code can be programmed into the Arduino microcontroller through the USB serial port. At this point the Arduino can be used as a PC or Raspberry Pi server peripheral hardware. Connecting the Arduino to the computer through USB can realize serial communication, and carry out the subsequent printing experiments and data acquisition work. Finally, the four sensors, including necessary components such as crystal oscillators, are integrated into the Arduino microcontroller for unified coordination and management. In addition, we format the data in the Arduino microcontroller controller. On the PC side, the sensor and time data in the Arduino microcontroller controller can be read synchronously through the serial debugging assistant. These data can be written to the PC through the serial port for subsequent analysis.

5.4   Conclusion

This chapter mainly discusses the different 3D printer classification methods, and the two main 3D printer access methods, including adapter-based access and sensor-based access. In the adapter-based access, this chapter proposes the technical framework of 3D printing online adaptation access and service, and then specifically describes the implementation method of the online adaptation access technology, which mainly includes three aspects. First, the functional design of the adaptive access device is given, which supports data collection and online operation of different types of devices and multiple devices. The second is to provide the platform-side functional design to support the management of access devices and 3D printing equipment. The third is to provide the communication mechanism between the adaptation access device and the cloud platform, which supports online processing task invocation and real-time data monitoring and control. In addition, we discussed how to implement Distributed slicing task execution in the adapter-based 3D printer access mode. In the sensor-based access technology, the design of 3D printing data acquisition system is discussed, including the construction of related hardware and the design of data acquisition scheme. First, we selected some suitable sensors based on printing experience and prior knowledge, and placed these sensors on suitable 3D printer parts. Finally, the sensor is integrated into the Arduino microcontroller for unified management. Based on the built 3D printing data acquisition system, the experimental scheme is designed. Finally, the 3D printer status data required for subsequent analysis can be obtained through multiple printings.

References

[1] Wohlers Report 2020, 3D printing and additive manufacturing global state of the industry, Wohlers Associates, 2020.

[2] A. Gebhardt, J. Kessler, L. Thurn, 3D Printing: Understanding Additive Manufacturing, Hanser Publishers, 2019.

[3] A.S. Mark, M. Jochen, W.V. Claas, A.L. Jennifer, Voxelated soft matter via multi-material multinozzle 3D printing, Nature 575 (2019) 330–335.

[4] G. Galdos, P. Tamigniaux, J.P. Morel, et al., Vibration Sensor, 2017.

[5] M. Li, L. Zhao, The classification of human lower limb motion based on acceleration sensor, in: Proceedings of 2016 IEEE Chinese guidance, navigation and control conference (CGNCC), IEEE Xplore, 2017, pp. 2210–2214.

[6] L. Guerriero, G. Guerriero, G. Grelle, et al., Brief communication: a low-cost Ar-duino®-based wire extensometer for earth flow monitoring, Nat. Hazards Earth Syst. Sci. 17 (6) (2017) 881–885.

[7] A. Samourkasidis, I.N. Athanasiadis, A miniature data repository on a raspberry pi, Electronics 6 (1) (2017).

[8] L. Guerriero, G. Guerriero, G. Grelle, et al., Brief communication: a low-cost Arduino®-based wire extensometer for earth flow monitoring, Nat. Hazards Earth Syst. Sci. 17 (6) (2017) 881–885.

[9] S. Bose, S. Mukherjee, S. Kundu, et al., AVR microcontroller based conference presentation timer, in: Proceedings of the First International Conference on Intelligent Computing and Communication, Springer, 2017, pp. 407–412.

[10] N.S. Kumar, B. Vuayalakshmi, R.J. Prarthana, et al., IOT based smart garbage alert system using Arduino UNO, in: Proceedings of 2016 IEEE region 10 conference (TENCON), IEEE Xplore, 2017, pp. 1028–1034.

Chapter 6

3D printing process monitoring

6.1 Quality problems in the 3D printing

For current 3D printing technology, printing a high quality 3D printed product generally takes a few hours. Printing failure often occurs due to various faults in the printing process, which may result in interruption of the printing process or poor quality of the printed product. For example, the printing material is exhausted or broken, resulting in the uncomplete product (Fig. 6.1); at the start of the printing, the material does not adhere to the printing base firmly, resulting in the poor-quality product (Fig. 6.2).

If there some faults in the printing process, the product will be unable to use. And if the user fails to notice when the printing faults, time and materials will be wasted. If users want to keep it operating properly and discover the above-mentioned possible faults in time, they need to manually check 3D printer all the time. This will lead to unnecessary waste of time and manpower and therefore reduces the overall efficiency of successful printing. Especially, when users manage multiple 3D printers at the same time with cloud environment and distribution collaboration, manual supervision of the printer becomes even more infeasible. What we want is a detection method and corresponding devices that can monitor the status of the 3D printer in real time and find the fault as early as possible and provide an alarm service for the users [1], thereby saving a lot of time and printing materials.

In recent years, 3D printing technology has developed rapidly, especially in the industrial field and medical field. However, the technology of 3D printing fault detection still lags behind. A 3D printing product is printed layer by layer according to the slicing file of the 3D printing model. Therefore, as long as there is a 3D model, objects of various shapes can be printed at one time. This special process makes the methods used for pipeline product detection no longer applicable to the detection of 3D printing products.

At present, the detection methods of 3D printing product quality used in the industry mainly depends on computer vision. As early as 1982, Zuech Nello [2] pointed out that quality detection through machine vision would be an important development direction in the future, and machine vision could liberate human beings from tedious quality inspection work.

Customized Production Through 3D Printing in Cloud Manufacturing
https://doi.org/10.1016/B978-0-12-823501-0.00003-1

FIG. 6.1 The uncomplete product.

FIG. 6.2 Poor quality product.

In recent years, the research on fault monitoring of 3D printing has become more and more active, especially machine learning, machine vision, image processing and other technologies have been used for 3D printing fault monitoring [3,4]. Mohammad et al. [5] developed a CNN Deep Learning model to detect real-time malicious defects built on the concepts of image classification and computer vision using machine learning to prevent the production losses and reduce human involvement for quality checks. The main drawbacks are the inability to detect defects in the vertical plane. Shen et al. [6] provided a self-feature extraction method to distinguish the shape defect of 3D printing products. Liu et al. [7] proposed a multi-edge feature fusion algorithm to detect the defects on low-quality printing images. Delli et al. [8] proposed a machine learning method, support vector machine (SVM), to assess the quality of 3D printed parts with the integration of a camera, image processing, and supervised machine learning.

The methods based on machine vision will be subject to many limitations, such as lighting conditions, occlusion caused by printer structure and so on. This chapter introduces a method to predict printing faults through the changes of relevant parameters in the printing process of 3D a printer. This method can detect the most common printing faults for a commonly used 3D printer, and can be extended to different faults of other types of printers [9]. It is a supplement to the monitoring method based on machine vision.

6.2   Overview of the 3D printing process monitoring method

The purpose of the 3D printing process monitoring method in the cloud manufacturing environment is to detect the printing status using the data collected from multiple sensors. The problem to be solved is essentially a classification problem. The machine learning method is used to train the data off-line and generate the classifier model. Finally, the model is uploaded to the cloud platform and invoked online as a cloud service. The method can be summarized as follows (Fig. 6.3):

1. **Data acquiring**

    Firstly, the printer parameters that may be related to the health state are selected according to the prior knowledge, and then do experiments with a 3D printer of the same model as the printers under study. Through experiments, a set of data sets characterized by key parameters of 3D printer and 3D printing fault status as quantitative label are obtained. The data set will be used for training the classification models. To carry out the data acquiring, there are three things need to be done, including hardware design, sensor selection and experiment method.

2. **Data preprocessing and analyzing**

    The original data gotten from hardware is not usable directly. After getting the raw data of the sensors, in order to train the data needed by the

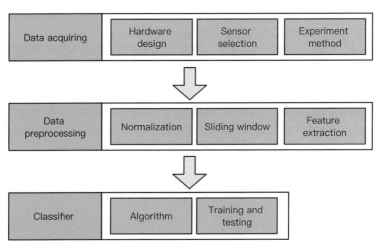

FIG. 6.3   The framework of 3D printing process monitoring method.

classifier, they need to be preprocessed and analyzed, including normalization, sliding window and feature extraction.

3. Classifier

Then we need to choose the proper algorithm to train the classifier. The data obtained by the above method is used to train the classifier and the classifier will be tested to evaluate the classification effect.

This chapter introduces the idea and implementation process of this method by taking a Fused Deposition Modeling (FDM) printer as an example.

6.3 3D printing fault detection based on process data

6.3.1 Data acquiring

1. The choice and installment of sensors

In order acquire data, we need to select the appropriate sensors and install them in the appropriate positions of a 3D printer. On the one hand, the sensors are used to obtain the 3D printer's relevant parameters that will be provided for subsequent analysis and the training of fault detection models. On the other hand, real-time data will be obtained from sensors and uploaded to the cloud for real-time monitoring.

It is very challenging to select appropriate sensors to have a good monitoring effect. What kind of sensors to choose depends on the structure of a 3D printer and the types of faults we want to monitor. At present, there is no general method to choose sensors. Prior knowledge and printing experiences can be used for making an initial choice. The final choice needs the help of experiments. Take the FDM (Fused Deposition Modeling) printer (Fig. 6.4) as an example, the movement of the printer nozzle is composed of three parts. The X and Y axes move horizontally, while the Z axis moves vertically. The movement in the three directions is driven by stepping motors on the three axes respectively.

According to the above principle, the most important parts to control the motion of the printer are the nozzle and the three stepping motors. Therefore, it is considered to place sensors for these four parts. Two vibration sensors are installed on x-axis and y-axis stepper motors. No sensors are installed on the z-axis stepper motor, which is inside the body of the 3D printer and is difficult to install. A vibration sensor and an acceleration sensor are installed on the nozzle component, in which the vibration sensor can also detect the vibration signal in the z-axis direction. Six state parameters of 3D printer will be obtained through the four sensors,. The values of the six state parameters will be recorded at each sampling time. The output of each sensor will be integrated into Arduino or Raspberry Pi server peripheral hardware [10,11]. The details on the use of sensors can be found in Chapter 5.

In fact, in addition to vibration and acceleration sensors, we have investigated more sensors including current, voltage, temperature and humidity.

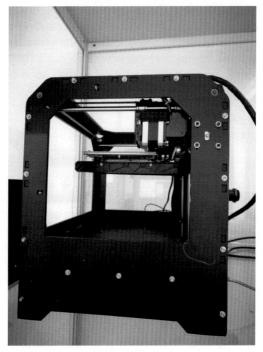

FIG. 6.4   FDM (fused deposition modeling) 3D printing.

Through repeated experiments, we finally chose vibration and acceleration sensors. It should be noted here that not the more sensors are installed, the more accurate the prediction results will be. The selection of sensors is still one of the most difficult problems in fault detections of 3D printing.

2. **Acquisition of data samples**

In order to analyze the data, we need to get a large labeled sample set. Each sample takes the six-dimensional data of multiple sensors as the eigenvalue and labeled with fault (0) or normal (1).

The data recorded by the sensor is a numerical matrix composed of time-series eigenvalues, and the label set must correspond to it one by one through time. Therefore, in the process of collecting printing data through printing experiments, the 3D model must be regular objects with exactly the same horizontal cross-section, such as cylinders and cuboids.

Therefore, during the experiments, the cross section of the printing product is controlled to be the same, hence the printing time of the current height can be obtained according to the height, so that a certain time corresponds to a certain height. Based on this method, different cross sections are printed many times and multiple groups of data are generated as a sample set for analysis.

For example, in the case of curling caused by not sticking to the printing at the initial stage of printing, the time can be calculated according to the printing height, so as to correspond to the collected sensor data to obtain the data with label; For the case that consumables are broken during printing, the corresponding data can be marked directly according to the height of the printed product.

6.3.2 Data preprocessing

The original data obtained in the previous section is shown in Fig. 6.5. There are the following problems:

(1) The sensitivity to noise is relatively high. This is because that 3D printer produces relatively large vibrations at the edges of the 3D product. This is not necessary for data analysis, and it is not easy for visualization and data understanding when they exist, therefore, they can be treated as noises.

(2) Six dimensional parameters have different dimensions and the sensor will produce signals in different numerical ranges when different products are printed. As a result, data obtained from different sensors and different printing will not be comparable. The data set needs to be normalized.

(3) We actually need to know whether the printer is in an abnormal state in a small period of time (an abnormal state at a certain moment may encounter noise interference), so the data can be smoothed to get a better prediction.

Therefore, in the preprocessing stage, the following two basic processes should be carried out, that is, the time window processing and standardization processing.

There are three methods of time window processing: fixed time window, recursive window and rolling window. We use the rolling window method to process the data obtained in last section. In this method, the length of the time window is fixed, while the training samples will be adjusted.

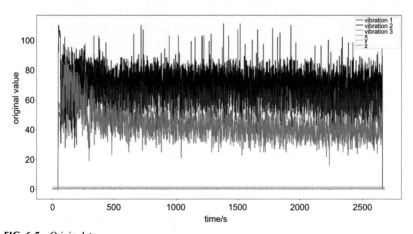

FIG. 6.5 Origin data.

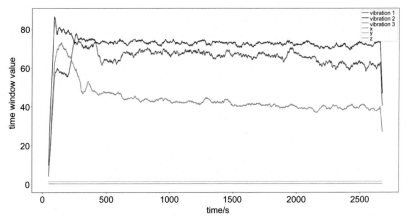

FIG. 6.6 Data after preprocess of time window.

Taking every n data points of time series data as a time window, the sensitivity to noise points can be reduced by selecting the appropriate value of n. The determination of n value needs to be tried repeatedly in experiments. Fig. 6.6 shows the data after the preprocessing of sliding window ($n = 50$).

Data normalization is to scale the data to a specific interval, so that the numerical range of multiple printing data can be controlled within one range. Thus, the features between different dimensions can be compared numerically, which can greatly improve the accuracy of the classifier. The normalization of data in deep learning can also prevent the gradient explosion of the model. The normalization methods mainly include min max method and Z-score method. Through the comparative test, we use the Z-score method, which can significantly improve the classification accuracy of the model. Fig. 6.7 shows the data after z-score normalization.

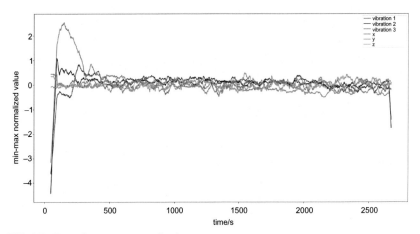

FIG. 6.7 Data after z-score normalization.

6.3.3 Feature engineering

In the field of machine learning, before the actual analysis of data, in addition to data preprocessing, feature engineering should be carried out to extract the most effective features for the problem to be studied. After data preprocessing, we get six-dimensional parameter data and one-dimensional label data.

The extracted features are used for models and algorithms. The following feature extraction methods are used.

(1) In order to comprehensively consider the parameters of the three vibration sensors and the parameters of the acceleration sensor, their respective sums are taken as one-dimensional features.
(2) Considering that the parameters of vibration sensor and acceleration sensor in the same direction may be correlative, the original six-dimensional data are combined in pairs to get better classification effect.

After the above feature extraction, the AUC of the model is improved by about 2%.

6.3.4 Classifier

Two kinds of faults will be studied here: curling in the initial stage of printing and material breaking in the middle of printing. For these two kinds of faults, classifier training is carried out respectively. The training methods for the two faults are the same, but there are slight differences in the data labeling process.

1. The choice of the classification models

Since the two kinds of faults can be regarded as a two-class problem (whether it is in a fault state) and the number of features is not large, it is preferable to use the LR algorithm [12], while SVM and Random Forest algorithm are used for comparisons.

Logistic regression is based on the linear regression model.

Linear regression is

$$z = \theta_0 + \theta_1 x_1 + \theta_2 x_2 + ... + \theta_n x_n = \theta^T x \qquad (6.1)$$

Logistic regression model is

$$h_\theta(x) = \frac{1}{1 + e^{-z}} = \frac{1}{1 + e^{-\theta^T x}} \qquad (6.2)$$

The output of the logistic regression model is a probability, between (0,1). If $h_\theta(x) < 0.5$, it belongs to 0 category; If $h_\theta(x) > 0.5$, belongs to 1 category.

$$P(y = 1|x; \theta) = h_\theta(x) \qquad (6.3)$$

$$P(y = 0|x; \theta) = 1 - h_\theta(x) \qquad (6.4)$$

The corresponding probability function is

$$P(y|x;\theta) = (h_\theta(x))^y * (1 - h_\theta(x))^{1-y} \tag{6.5}$$

The likelihood function corresponding to m samples is

$$L(\theta) = \prod_{i=1}^{m} \left(h_\theta\left(x^{(i)}\right) \right)^{y^{(i)}} * \left(1 - h_\theta\left(x^{(i)}\right) \right)^{1-y^{(i)}} \tag{6.6}$$

The log likelihood function is

$$l(\theta) = \log\left(L(\theta)\right) = \sum_{i=1}^{m} y^{(i)} \log\left(h_\theta\left(x^{(i)}\right) \right) + \left(1 - y^{(i)}\right) \log\left(1 - h_\theta\left(x^{(i)}\right) \right) \tag{6.7}$$

Finally

$$J(\theta) = -\frac{1}{m} l(\theta) \tag{6.8}$$

Then the gradient descent method can then be used to find the optimal θ parameter.

Random Forests [13] are based on decision trees and use bagging methods to integrate learning. Support vector machine [14] uses the theory of maximizing the interval.

2. Training of the classifier

According to the above description of the experimental method, the quality of the printed product can be judged by the appearance, and it can be determined whether the printed product is normal on a certain height, and then according to the correspondence of time and height, we can know the label of a certain data, fault or normal. In this way, we can get labeled data.

After data of each printing is preprocessed, a data set is obtained. Combine multiple labeled data from different printing as an entire data set, and then we can start model training.

First, divide the dataset obtained above into a training set and a test set. The test set data does not participate in the model training process and is only used to test the classification accuracy of the model.

The classification accuracy refers to the correct number of test sets classified/the number of test sets.

The evaluation indexes of the model used here are accuracy, AUC score, recall score, precision score, where the principal index is the AUC score. Performance comparison on poor quality product problem is shown in Table 6.1. From the table we can see that:

(1) All the three algorithms can achieve a classification accuracy of more than 90%. LR and SVM AUC algorithms can achieve AUC score of more than 90%.

TABLE 6.1 Performance comparison on poor quality product problem.

	LR	SVM	Random forest
Accuracy (%)	91.76	91.56	90.54
AUC score (%)	90.00	90.78	87.58
Recall score (%)	87.09	89.49	82.59
Precision score (%)	75.40	72.56	73.39
Time (s)	55.00	55.44	238.66

(2) The running time of the Random Forest is relatively too long.

(3) The LR is better than SVM in time and precision score.

(4) Considering the simplicity and interpretability, we'll choose LR as the final model.

Performance comparison on uncomplete product problem is shown in Table 6.2. From this we can see that:

(1) All the three algorithms can achieve accuracy and AUC of more than 90.

(2) The Random Forest algorithm is better than the others after parameter tuning.

(3) The Random Forest algorithm is slightly insufficient in time.

(4) For a comprehensive assessment, the Random forest can be used as a model in the final system.

3. Fusion of classifiers

Several models have been tested in the above section. Because different models are based on different principles, the classification effects of different models are different. This difference can be used to make the prediction results more accurate through the combination or fusion of models. The higher the accuracy and diversity of individual models, the better the fusion effect.

TABLE 6.2 Performance comparison on uncomplete product problem.

	LR	SVM	Random forest
Accuracy (%)	92.5	94.93	95.43
AUC score (%)	96.5	95.93	96.61
Recall score (%)	91.35	92.79	93.64
Precision score (%)	93.67	96.96	97.52
Time (s)	28.31	32.03	36.83

There are several ways of classifier fusion as follows:

(1) Simple methods

The fused results are obtained directly according to the prediction results of different classifiers. In this way, there is no need to train another classification model or retrain the model. Only the prediction results of several basic models can be obtained. The frequently used methods include majority voting fusion, weighted voting fusion, weighted average to the results, etc.

(2) Learning methods

The learning method is to take the individual classification model as the primary one, and then train another model to fuse multiple different models. The most common learning methods include Stacking and Blending.

The core idea of Stacking is to use one or more basic classifiers to learn from the initial training set, train the primary classifiers, take the learning results obtained by these primary classifiers as features, and then use another classifier to train and integrate the new features [15]. Through Stacking, the performance of different models obtained from several training can be integrated, and finally the effect of the whole classifier can be improved.

Blending and stacking are similar [16], and their principles are roughly the same. The difference is that Blending's training set does not obtain the predicted value through k-fold's cross validation (CV) strategy to generate the characteristics of the second stage model, but establishes a Holdout set, that is, Blending uses the Holdout method instead of the n-fold method when obtaining the prediction results in the first layer. Blending is simpler than Stacking, but there may be overfitting. Stacking makes full use of the whole training data, so the training will be more stable. In order to get more stable classification results, we use stacking to fuse the classifiers. The overall effect of the fused classification models is shown in Table 6.3.

TABLE 6.3 The effect of classifier fusion.

	Curling problem in the initial stage of printing	Material breaking problem in the middle of printing
Accuracy (%)	92.15	96.43
AUC score (%)	91.33	96.90
Recall score (%)	83.23	94.78
Precision score (%)	87.27	98.35

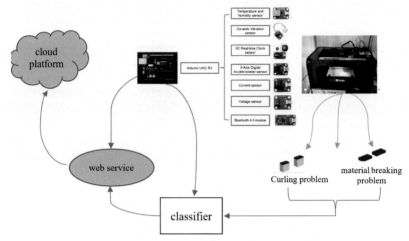

FIG. 6.8 Construction of the cloud service and connection to cloud platform.

After the model fusion, the required complete model is obtained. When a new sample is given, the probability value of the current fault state of the sample can be calculated by the fault model. A reasonable threshold can be set according to the user's sensitivity to early warning. When the threshold is large, the precision of the policy will be high and the recall will be low. On the contrary, the recall and precision will be larger. According to statistical experimental results, when the threshold is set at 0.85–0.9, The classification effect will be very good.

6.3.5 Design of the fault detection system

After establishing the 3D printing fault detection model, it is necessary to build a cloud service on the server to obtain the real-time data of the printing process of each 3D printer, and predict the quality by combining with the quality detection model obtained in the previous part. Finally, the service is connected to the cloud 3D printing platform for use, as shown in Fig. 6.8.

6.4 Conclusion

Based the introduced methodology, the status of the 3D printing can be monitored by several sensors, and the machine learning method can be used to predict whether the printer is in a fault state. Specifically, based on experiments we have done, the vibration of the stepping motor in the 3D printer is most important for the failure detection. Therefore, the 3D printer provider can add the built-in vibration sensor in the stepping motor at the factory, so that the real-time fault can be monitored.

Of course, the classification model given in this chapter is not applicable to all printers and fault types. However, based on the idea of this chapter, real-time fault detection solutions for more 3D printers can be developed. The main problem of this methodology is the strong dependence on data sets.

In the future, two kinds of research can be conducted. One is to establish labeled data sets for different types of printers and faults, which will need a huge amount of work. The other research is to study unsupervised, semi supervised, or reinforcement learning methods to reduce the dependence on labeled data, which is a promising research direction.

References

[1] X. Liu, Research on Establishment of Internet-based Product Quality Tracking System in Mobile Internet Age, Standard Science, 2016.

[2] Z. Nello, Machine vision and quality control, Laser Institute of America, 33 (1982) 21–23.

[3] X. Qi, G. Chen, Y. Li, X. Cheng, C. Li, Applying neural-network-based machine learning to additive manufacturing: current applications, challenges, and future perspectives, Engineering 5 (4) (2019) 721–729.

[4] L. Meng, B. McWilliams, W. Jarosinski, H.Y. Park, Y.G. Jung, J. Lee, J. Zhang, Machine learning in additive manufacturing: a review, JOM 72 (6) (2020) 2363–2377.

[5] F.K. Mohammad, A. Aftaab, A.S. Mohammad, S.A. Mohammad, R. Yasser, S. Nehal, A.S. Ibrahim, Real-time defect detection in 3D printing using machine learning, Mater. Today Proc. 42 (Part 2) (2021) 521–528.

[6] H.Y. Shen, W.Z. Du, W.J. Sun, et al., Visual detection of surface defects based on self-feature comparison in robot 3-D printing, Appl. Sci. 10 (2020) 235, https://doi.org/10.3390/app10010235.

[7] B. Liu, Y. Chen, J. Xie, B. Chen, Industrial printing image defect detection using multi-edge feature fusion algorithm, Complexity (2021) 2036466.

[8] U. Delli, S. Chang, Automated process monitoring in 3D printing using supervised machine learning, Procedia Manuf. 26 (2018) 865–870.

[9] B. Li, L. Zhang, L. Ren, X. Luo, 3D printing fault detection based on process data, in: Proceedings of 2018 Chinese Intelligent Systems Conference. Singapore, Springer, 2019, pp. 385–396.

[10] L. Guerriero, G. Guerriero, G. Grelle, et al., Brief communication: a low-cost Arduino®-based wire extensometer for earth flow monitoring, Nat. Hazards Earth Syst. Sci. 17 (6) (2017) 881–885.

[11] A. Samourkasidis, I.N. Athanasiadis, A miniature data repository on a Raspberry Pi, Electronics 6 (1) (2017).

[12] S. Edition, Applied logistic regression analysis, Technometrics 38 (2) (2017) 184–186.

[13] S.J. Rigatti, Random forest, J. Insur. Med. (2017) 31–39.

[14] R. Fu, B. Li, Y. Gao, et al., Content-based image retrieval based on CNN and SVM, in: *Proceedings of 2nd IEEE International Conference on Computer and Communications*, Chengdu, China, IEEE Xplore, 2017, pp. 638–642.

[15] A. Alves, Stacking machine learning classifiers to identify Higgs bosons at the LHC, J. Instrum. 12 (5) (2017).

[16] K. Lattery, A machine learning approach to conjoint analysis: boosting and blending ensembles, in: Proceedings of Sawtooth Software, 2015, pp. 353–378.

Chapter 7

3D printing credibility evaluation

As described in Chapter 3, although there are few steps in 3D printing process, each link needs to be accurate in order to finally print qualified products. In the 3D printing cloud platform, the functions, models, and devices in Fig. 3.1 are in the form of services. Every service needs to be strictly verified and evaluated. This chapter will introduce 3D printing service evaluation methods by taking the two typical services of 3D printing model design and 3D printing as examples.

7.1 3D printing model credibility evaluation

7.1.1 Problem description

Designing 3D printed models is a highly specialized work. A well-designed 3D printing model cannot be used directly for printing. It needs to be sliced and parametrically designed for the characteristics of the 3D printer, such as printing accuracy, printing materials, etc. These parameters will restrict the quality of the final part. After the slicing parameters are set, the printing parameters are designed. This step needs to set a series of parameters, including part position placement, part size scaling, layer filling rate, support position selection, etc. Even two 3D printers of the same model also require minor adjustments of parameters due to differences in actual printing performance.

In the 3D printing environment of industrial mass production, the above design parameters involve different types of workers. For example, in the production process of 3D printed parts for BMW cars, the above design parameters involved at least four different types of personnel, including part geometry designers, slice parameter process designers, on-site 3D printing operators, and material performance analysts. It can be seen that 3D printing model design is actually a huge design project. There are many opportunities for mistakes in this process. Any error or error in any parameter will result in the failure of the print job. Therefore, it is difficult to directly print a usable part until these model design parameters have not been verified.

Here we give evaluation parameters for the design parameters that may lead to printing failure in the design process of 3D printing model. On this basis, the

Customized Production Through 3D Printing in Cloud Manufacturing
https://doi.org/10.1016/B978-0-12-823501-0.00005-5

design parameter evaluation index is verified by using 3D printing design software with parameter verification function, so as to improve the overall success rate of 3D printing. In terms of technical implementation, the relevant 3D printing design software can be migrated to the cloud platform by using lightweight virtualization technology. In this way, a service component with parameter validation function can be formed. These service components provide a verification environment for the plausibility assessment of 3D printing models.

7.1.2 Framework of 3D printing model credibility evaluation

The implementation framework for 3D printing model credibility evaluation is shown in Fig. 7.1. First, according to the relevant design parameters in the model library, analyze the characteristic parameters that may affect its printing success rate. The domain meta-model is constructed according to the model description specification of the domain to which these feature parameters belong. On the one hand, XML technology is used to configure the meta-model into the visual interface of the front-end web page of the service component. On the other hand, the parameters of these metamodels are verified and solved separately based on domain simulation software. For example, ANSYS Additive is used to solve the mechanical and thermal performance parameters of the printing material, so as to avoid the failure of part deformation or edge curling due to improper setting of the heating temperature of the 3D printer base during the actual printing process. For example, using Autodesk Netfabb design verification software can analyze the closedness of the polygonal geometry of the part model. At the same time, Autodesk Netfabb can be used to analyze the configuration of the center of gravity of the geometric model of the part and the space ratio of the support occupied by the whole part, so as to avoid the collapse of the part due to unbalanced force during the printing process. For another example, using Simplify3D can simulate the actual operation of the nozzle path of a 3D printer, so that a very accurate part printing time can be calculated.

In the verification and solution process of the domain metamodel, the domain simulation software is driven by using lightweight virtualization technology and script programs, and the solution results of the simulation software are loaded into the model parameter configuration interface on the front end of the service component web page. For the credibility judgment of the parameter solution results, we need to delineate the effective range of these parameters to evaluate the quality of the results. The selection of the effective range can be based on the extraction and mining of the tacit feature knowledge in the expert knowledge base to determine the effective threshold range of the model design credibility index. The parameter evaluation method based on the graphics principle in the field simulation software can also be used to guide the effective selection of parameters.

Finally, based on the simulation verification, the design parameters verified by the simulation are loaded into the 3D printed slicing software for physical

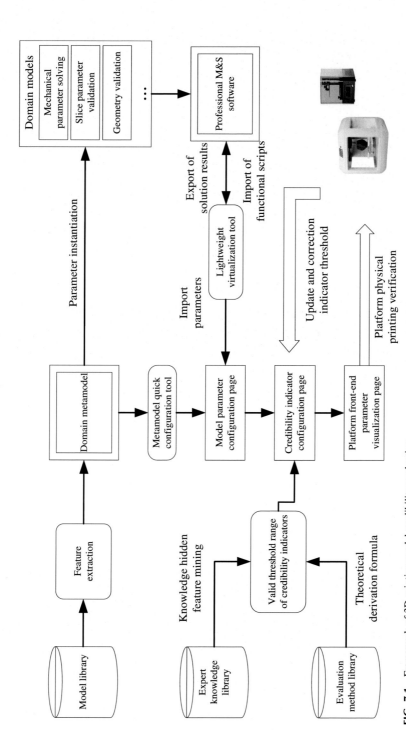

FIG. 7.1 Framework of 3D printing model credibility evaluation.

testing. Then the actual printing effect is used as a case sample of the configuration range of the credibility index. In this way, the valid threshold range of the parameters can be further revised.

Field simulation software for 3D printing is usually bulky. For example, the process of finite element analysis of 3D printing material properties using ANSYS would be computationally intensive. It is obviously uneconomical to configure such computing tasks into distributed server nodes in a cloud platform. Therefore, we can use lightweight virtualization technology, that is, through a scripting language to enable remote clients to communicate with servers with domain emulation software installed. Then we load the solution results into the visual interface of the relevant service components in the front-end of the platform web page. In this way, only the domain simulation software needs to be deployed in a well-configured cloud server, and the client nodes can call the computing resources of the server to solve the relevant parameters.

7.1.3 Indexes and evaluation of 3D printing model credibility

This section will analyze in detail the credibility indicators of several important attributes that directly affect the printing quality in the 3D printing model, and analyze the credibility range and evaluation methods of these indicators.

1. **The credibility assessment of the geometric topology of the 3D mesh of a part**

    From the knowledge of calculus, it can be known that if you want to find the circumference π of a circle, you can divide the circle into N inscribed polygons. When the number of N is larger, the obtained approximate perimeter is approximately close to π. 3D printing subdivides a parts based on this knowledge. The geometry of a part is usually subdivided using triangles or polygons. These subdivided triangles or polygons form the geometric topology of the part's 3D network. Commonly used digital representations of these geometric topologies are in STL and OBJ formats. The 3D printed slice file is a collection of closed annular figures obtained by horizontally slicing the geometric topology of the part. However, the geometric topology of complex parts is not very regular. In the process of slicing the geometric topology of complex parts, problems such as overlapping triangular patches, splicing cracks, and internal voids may occur. These issues can result in incomplete slices of geometry, preventing them from printing properly. These problems will gradually increase as the slice layer thickness decreases. 3D Systems has standardized the STL format for the above problems, and has formulated consistency rules, so that the above errors can be easily detected. At present, the more mature 3D printing slicing software has built-in related algorithms to automatically fix these errors. But the

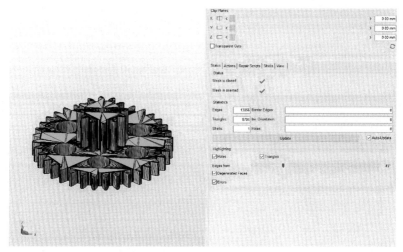

FIG. 7.2 Example of credibility assessment of 3D mesh closure.

automatic repair process inevitably changes the original design shape of the part, resulting in varying degrees of distortion in the final printed product.

Therefore, the distortion rate of the edge of the actual printed product can be considered as the basis for judging the credibility of the 3D network geometric topology of a part. The credibility index can be defined as the ratio of the total number of triangles in the STL file to the number of automatically repaired triangles after running the 3D slicing software. And these two parameters can be read from 3D slice software. Through statistical analysis, the proportion of parts with errors is generally one-tenth to one-seventh. If the error ratio is higher than this range or fails to pass the algorithm self-check, it is considered that the geometric design of the part is not satisfactory and needs to be redesigned. The mesh closure of parts can be verified using Autodesk Netfabb software. We take the gearbox pinion model as an example to verify the mesh closure, as shown in Fig. 7.2. The gearbox pinion is an industrial-grade 3D printed product. The 3D printed mesh of its parts has 8704 small triangles. It has been verified that all its triangles satisfy the mesh closure verification. Hence, there will be no edge distortion in printing.

2. **Credibility assessment of parts balance**

According to gravity, if a person wants to maintain a stable standing posture, the center of gravity of the body needs to be kept within a certain range with the support foot as the center. If the body is tilted to a certain extent, causing the center of gravity to shift, it will definitely not be able to stand stably. Similarly, when designing some parts with a camber angle, the position of the center of gravity must be considered. For a part with a camber angle, when its center of gravity falls within a certain range of its supporting

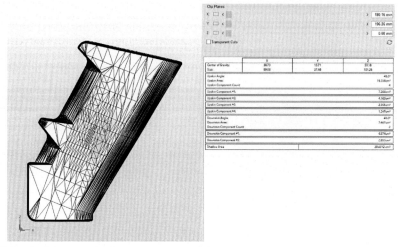

FIG. 7.3   Balance evaluation based on the Netfabb software.

surface, the balance of its part can be maintained. No additional support is required in this case. When its center of gravity falls outside a certain range of its supporting surface, it is often impossible to maintain the balance of a part. Therefore, it can be considered that the parameters affecting the credibility of balance are the position of the center of gravity of the part and the extent of its supporting surface. The evaluation index is whether the center of gravity falls within the range of 0–140% of its support surface. The center of gravity position and support surface area of the model can be solved automatically using Autodesk Netfabb software. And based on this, the balance of the parts can be determined. This also provides a good reference for the evaluation of the parameter design of the part model. Some parts are designed with a large camber angle that cannot maintain the balance of the parts, but it is necessary to consider the actual position of the center of gravity to judge whether it is balanced. The part in Fig. 7.3 is the base part of a product. Although its appearance seems unbalanced. However, through the balance judgment based on the position of the center of gravity, the center of gravity actually falls at 102% of the area of the support surface. Indicates that the part is balanced. The green point in Fig. 7.3 is the center of gravity.

3. **Credibility assessment of additional supports**

Due to the layer-by-layer processing method in the 3D printing process, if the angle between the protruding part and the main part of the part is too large, the slicing software will automatically generate additional supports during the printing process to ensure normal accumulation.

As shown in Fig. 7.4, the additional supports will occupy a large part of the original model volume due to its too large tilt angle when printing based

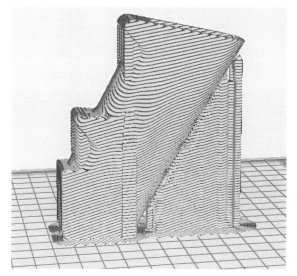

FIG. 7.4 Additional support in original position.

on the original placement of the base model above. As a result, the actual printing effect is not ideal. Therefore, when designing the part model, the inclination angle of its protruding part needs to be considered. Usually, 3D printers do not need to add support by default when the inclination angle of the protruding part relative to the main body is less than 40 degrees.

In order to evaluate the impact of supports on actual 3D printing, the ratio between the volume of the part and the support volume can be used as a credibility evaluation index for the additional supports. The reference range of this indicator is 0–5%. When it exceeds this range, it can be considered that the volume ratio of the additional support is too large. In this case, the design parameters of the support will affect the actual printing effect of the part. However, an experienced operator will change the placement of the part during actual printing, so that the inclined surface of the part is printed as the base datum. In this way, the volume ratio of the additional support of the part becomes 0.3%, which is within the range of the support credibility evaluation. Although a small amount of additional support is generated in this way, the actual printing effect of the part will not be affected by post-processing, as shown in Fig. 7.5.

7.2 3D printing service credibility evaluation

When there are a large number of printing services on the 3D printing cloud platform, how to choose a suitable printing service for a specific printing task becomes an important technique. The credibility evaluation of 3D printing services is an important reference when we select 3D printing services for 3D

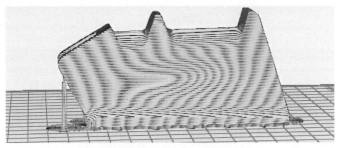

FIG. 7.5 Additional support after changing the print position.

printing tasks. There is no essential difference between the credibility evaluation of 3D printing services and the credibility evaluation of other types of cloud manufacturing services in terms of performance. There have been many studies on the credibility assessment of cloud manufacturing services. These methods can be used as references in the credibility assessment of 3D printing services.

7.2.1 Credibility evaluation indicators of cloud manufacturing services

The evaluation indicators of cloud manufacturing service credibility mainly include three aspects: business qualifications, technical capabilities, and service interaction.

1. **Business qualification evaluation**

    Business qualification evaluation mainly evaluates the qualification certification, operation management and financial aspects of the entities providing cloud manufacturing services from the perspective of the cloud manufacturing service platform. Among them, qualification certification includes platform certification and third-party certification. Operation management mainly evaluates the provider's market share and operational capability. Financial status mainly evaluates the status of assets and profits.

2. **Technical capability evaluation**

    Technical capability evaluation aims to objectively measure the status of cloud manufacturing service itself and its service products, mainly including technical level evaluation and quality level evaluation. The level of technology reflects a provider's ability to provide services, which can be evaluated through reliability, security, scalability, and specific metrics related to the manufacturing services. Reliability reflects the ability of cloud manufacturing services to complete their service content with stable performance. Security reflects the ability of cloud manufacturing services to ensure the security of users' physical resources and information resources. Scalability reflects the ability of cloud manufacturing services to handle large-scale concurrent requests. Specific metrics will be defined according to attributes of a given manufacturing service, such as a 3D printing service.

Quality level evaluation aims to measure the quality of cloud manufacturing services, which can be evaluated through accuracy, cost-effectiveness, and timeliness. Accuracy reflects the ability of cloud manufacturing services to meet user needs. The price-performance ratio reflects the matching degree between the fees paid for invoking cloud manufacturing services and the corresponding service products. Timeliness measures the time it takes for a cloud manufacturing service to complete its service content.

3. Service interaction evaluation

Service interaction evaluation aims to measure the interaction level between cloud manufacturing services and users from a subjective perspective, and objectively measure the logistics and transportation quality of service products. Interaction capabilities reflect the availability and ease of use of cloud manufacturing services. Availability reflects whether the cloud manufacturing service can provide services normally. Ease of use reflects the degree to which cloud manufacturing services are easy to be understood by service users and to implement service invocations. Logistics transportation mainly evaluates the transportation efficiency and logistics safety of service products. Transportation efficiency reflects the timeliness of service products delivered to service users through logistics channels. Logistics security reflects the quality assurance of service products in the process of logistics and transportation.

The indicators given above are applicable to the credibility evaluation of any kinds of cloud manufacturing services. For specific manufacturing resources and manufacturing capabilities, such as 3D printing services on a cloud manufacturing platform, specific evaluation metrics need to be designed based on their functional and performance attributes. Attributes of a 3D printing equipment are introduced in the following section.

7.2.2   Attributes of a 3D printing equipment

The attributes of a 3D printing equipment can be divided into static properties and dynamic properties, as shown in Fig. 7.6. The static attributes mainly include the basic information of the equipment, such as the size of the construction space, processing accuracy, equipment type, geographic location, etc. The parameters of these static properties of 3D printers generally remain unchanged. Dynamic attributes mainly include device running status and expected idle time. The parameters of dynamic service attributions of 3D printers usually change as the device operates.

1. Static properties

Corresponding to the processing attributes required by users, the basic attributes of 3D printing services reflect the capabilities of 3D printing equipment, including space size, material type, material color, processing accuracy, processing speed, equipment type, geographic location, etc.

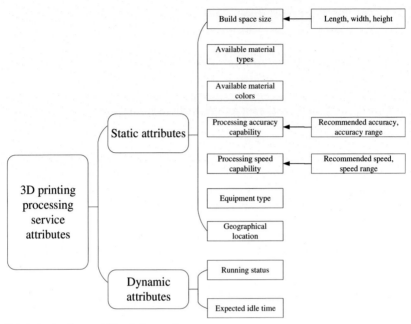

FIG. 7.6 Attributes of 3D printing services.

(1) Numerical type: Some attributions are represented by numerical values such as spatial size, machining accuracy, and machining speed. The spatial size includes the values of the three dimensions of length L, width W, and height H. The printing accuracy includes the recommended accuracy values and accuracy ranges. The recommended accuracy value is expressed as a numerical value, while the accuracy range is described as an interval. The processing speed includes the recommended speed values and speed ranges. The recommended speed value is expressed as a numerical value, while the speed range is described as an interval.

(2) Enumeration type: Material types, material colors, equipment types, and geographic locations are represented by enumeration values. The value range is the same as the value range of the corresponding attributions in the 3D printing task information. Note that there is a special value NEG in the value ranges of the basic 3D printing service attributes, which means "negotiable."

2. **Dynamic properties**

The dynamic properties of a 3D printing service mainly include two aspects: the running status and the expected idle time of the 3D printer.

(1) Running status

The online 3D printer running status can be described as idle, working, pause, stop, close, empty and error. The idle state means that there is no machining task currently. The working state indicates that the machining task is being executed. The paused state means that the currently executing task is paused. The stop state means that the current equipment is stopped but maintains data communication, such as startup maintenance, refueling, etc. The off state means the device is powered off and disconnected from data communication.

(2) Expected idle time

Assume that there are N tasks in a certain 3D printer task queue. The first task in the queue is the currently executing task, and its processing progress is p. The estimated printing time of the i-th task can be expressed as Te_i. The expected idle time T_{idle} of a device can be expressed as:

$$T_{idle} = Te_1 * (1 - p) + \sum_{i=2}^{N} Te_i \tag{7.1}$$

7.2.3 Credibility assessment process of cloud manufacturing services

The credibility assessment process of cloud manufacturing services is shown in Fig. 7.7 including service certification, service calls, service quality monitoring, and service evaluation [1].

1. Service certification

The cloud manufacturing service provides services to users based on the cloud manufacturing service platform. Users obtain cloud manufacturing services through the cloud manufacturing service platform. Cloud manufacturing service platform is the foundation and bridge for users to interact with cloud manufacturing services. The cloud manufacturing service platform implements basic qualification evaluation and dynamic certification management for cloud manufacturing services connected to the

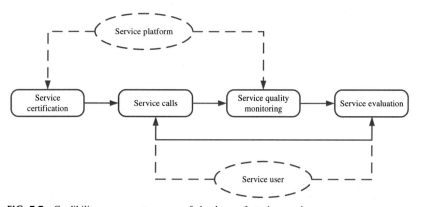

FIG. 7.7 Credibility assessment process of cloud manufacturing services.

platform. Generally, certified cloud manufacturing services are more trusted than uncertified cloud manufacturing services, and cloud manufacturing services with high certification levels are more trusted than cloud manufacturing services with low certification levels.

2. **Service call**

The historical data of cloud manufacturing service calls directly reflects the user's choice, and can also reflect the overall status of cloud manufacturing services over a period of time. Usually, on the premise of meeting basic service requirements, users are more inclined to choose cloud manufacturing services with good service reputation and high service quality. The historical data of cloud manufacturing service calls can more objectively reflect the credibility of cloud manufacturing services.

3. **Service quality monitoring**

The historical data of cloud manufacturing service quality monitoring is the objective data about the rationality and standardization of the service operation process and the service quality of each link recorded by the cloud manufacturing service platform during the service operation process. Cloud manufacturing service quality monitoring data directly reflects the service effect of cloud manufacturing services and users' satisfaction with cloud manufacturing services, and is an important objective basis for reflecting the credibility of cloud manufacturing services.

4. **User evaluation**

User evaluation data is an important basis for reflecting the credibility of cloud manufacturing services. The user evaluation data centrally reflects the degree of satisfaction of cloud manufacturing services to user needs, as well as the overall satisfaction degree of users with the service. User evaluation data is subjective. A small amount of user evaluation data is significantly affected by user subjectivity. However, a large number of user evaluation data can relatively objectively reflect the credibility of cloud manufacturing services because they cover a wider range of user groups and user needs. Cloud manufacturing service user evaluation data also has problems such as random evaluation, malicious evaluation, and data missing. Usually, statistical analysis or mining analysis and other technical means can be used to extract useful information from user evaluation data reasonably, accurately and efficiently to support the credibility assessment of cloud manufacturing services.

The evaluation methods are described in details in another book of the authors [2], which are not repeated here. When these methods are used for credible evaluation of 3D printing services, it is necessary to complete relevant steps and collect relevant data in the 3D printing cloud platform, including attributes of 3D printing equipment in the relevant indicators of technical capability evaluation.

7.3 Conclusion

3D printing service evaluation is an important guarantee for printing quality and printing success rate. In this chapter, we discussed credibility evaluation methods of two typical 3D printing services, 3D model design and 3D printing. For 3D model design, the framework for online evaluation of 3D printing model credibility on cloud platform is discussed. On this basis, the design parameters that affect printing quality and printing success rates are analyzed. Reference ranges and evaluation methods of their credibility are also proposed.

3D printing services can be taken as a typical cloud manufacturing service. We discussed the credibility evaluation of 3D printing services from the perspective of the cloud manufacturing service credibility evaluation. Service credibility indicators and the evaluation process for cloud manufacturing services are introduced. The functional attributes of the 3D printing services are described. Then existing methods of cloud manufacturing service credibility evaluation can be adopted to evaluate the credibility of 3D printing services by integrating the 3D printing service attributes. 3D printing service evaluation is the basis of 3D printing demand-provider matching and task scheduling that will be discussed in the next chapter.

References

[1] Z. Li, L. Liao, H. Leung, et al., Evaluating the credibility of cloud services, Comput. Electr. Eng. 58 (2017) 161–175.
[2] L. Zhang, Y.K. Liu, Service management and scheduling in cloud manufacturing, DE GRUYTER, 2022.

Chapter 8

Supply-demand matching and task scheduling

8.1 The supply and demand relationship in 3D printing cloud platform

3D printing is widely used in such sectors as industry, medical, sports and education with the rapid development 3D printing technology and continual breakthrough of new material technology. The survey carried out by 3D Hubs in April 2021 reveals that, as a result of the pandemic, 33% of engineering businesses increased their 3D printing usage. Faced with the continual expansion of 3D printing market and the diversity and rapid growth of the scale of 3D printing devices, more and more users order 3D printing services through cloud platforms. According to statistics, the average monthly orders of large commercial 3D printing platforms such as Shapeways and 3D-Hubs reached more than 200,000 pieces. How to quickly find suitable 3D printing services to efficiently finish a customized production task in cloud platform is a critical problem urgently to resolve. The problem is the typical supply-demand matching and task scheduling problem in cloud manufacturing environment, however, because of the characteristics of 3D printing devices, it is necessary to carry out specific research on the matching and scheduling of 3D printing devices in a cloud environment.

Order-oriented matching and production efficiency are key indicators that objectively reflect the platform's manufacturing capabilities. The capability of high efficient matching and scheduling of printing services with respect to an production order will be one of the core competitiveness of a 3D printing cloud manufacturing platform.

Through unified management of distributed 3D printing services of service providers, the cloud manufacturing platform achieves on-demand sharing of 3D printing services. The transaction model of cloud 3D printing services is shown in Fig. 8.1.

Demanders place their printing orders to the platform with desired 3D models. The 3D models are either submitted by the demanders or selected from the 3D model library in the platform. The 3D models then are sliced and

Customized Production Through 3D Printing in Cloud Manufacturing
https://doi.org/10.1016/B978-0-12-823501-0.00011-0
111

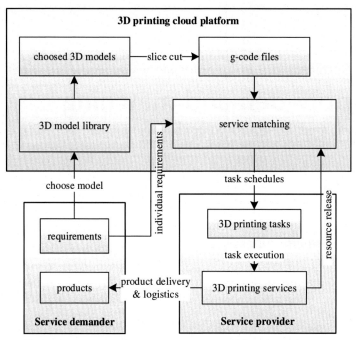

FIG. 8.1 The transaction model of 3D printing services in CMfg [1].

transformed to g-code files. 3D printing service matching solutions are generated through the service matching system according to the individualized requirements of customers, g-code file list and 3D printing service information. 3D printing tasks are allocated to the related services according to the generated service matching solutions. Then the 3D printing providers execute the received 3D printing tasks and deliver the printed products to the service demanders via the logistics.

Note that the geographically distributed 3D printing services of different service providers are different in processing capacity and attributes. Moreover, service matching system needs to allocate optimal 3D printing services for multiple tasks.

This chapter will discuss the 3D printing service supply-demand matching and scheduling problem under the transaction mode shown in Fig. 8.1.

8.2 Supply-demand matching

The 3D printing service matching problem is to choose suitable printing services to meet the demanders' requirements. Some research on 3D supply-demand matching in cloud environment has been conducted [2–6].

Nevertheless, systematic studies on the matching method for cloud 3D printing are far from enough. The current mainstream commercial 3D printing platforms are still in a relatively early online manufacturing stage. The services of such platforms fail to run through the production procedures in the whole life cycle of 3D printing products. In other words, the services of the platforms are independent of each other, without any series connection of service procedures between services. Moreover, since the dynamic information of printing devices and printing execution process is not fully utilized, the printing tasks may be failure.

This section presents a dynamic and static data based matching method for cloud 3D printing. A multi-source data integration model is established by considering the characteristics of the printing task and printing services in the cloud manufacturing environment. A capability indicator model is built based on bipartite graph to achieve a precise match. A standardized matching process and a set of matching rules are proposed to support the matching [7].

8.2.1 Multi-source data integration model

The multi-source data integration model is a normalized model that integrates the multi-dimensional data sources for cloud 3D printing through a unified description specification. A modeling framework is introduced to describe two models of the printing task and printing service and built a description standard for cloud 3D printing model. Therefore, from a technical implementation perspective, we adopt the concept of top-down for modeling builds a model description framework. This framework lays a foundation for quantitative matching of multi-dimensional 3D printing data sources. However, multi-source data integration model contains not only the information for matching but also the information for network communication and model retrieval (such as user's information and timestamp information, etc.). To realize the matching based on the bipartite graph, we pruned the redundant parameters of the model and formed the capability indicator model. A detailed description of the multi-source data model for printing tasks and printing services is as follows.

1. **Multi-source data integration model of printing tasks**
    The current research focuses on how to match the mechanical properties of 3D printers with printing tasks. For instance, the model of printing tasks is established such six dimensions as workpiece accuracy, workpiece surface roughness, tension, ductility, workpiece cost and building time. Based on the achievement of predecessors, we build a set of models for printing tasks applicable to the service-oriented distributed network manufacturing environment. As formula (8.1):

    $$T = \left(t_{basic-attribute}, t_{print-attribute} \right) \qquad (8.1)$$

    Print task models include two parts: the basic attribute model and printing attribute model. Since the application scenarios and system environments of models are different, the description dimensions and granularity of models are also different. In the service-oriented cloud manufacturing environment, modeling of printing tasks shall not be limited to the issue of matching with the parameters of adaptable 3D printers. It is also necessary to consider such communication problems as how to actualize the model sequencing, inquiring and matching in the network environment. Regarding various factors, modeling is carried out by the top-down design pattern, so that the models are universal and reusable.

Basic attribute model is a template provided by the system for users to create printing tasks. Platform users are required to fill up relevant information under the framework of the template before the system backstage program eventually transforms generally such input information into regular printing task description information. Basic attribute information includes user information, task description information, task ID and timestamp, as formula (8.2). User information includes order models quantity, delivery mode, user's name, user's address and other registration information input by the users when publishing the printing tasks. Task description information is the regular task description information generated automatically by the system, including task name, task description text, and user's evaluation result collected subsequently. Task ID is the serial number of task generated automatically by the system after the user fills in the user's basic information. The timestamp is the time record generated automatically after the user submits the task, used for the following task sequence and task allocation. Such basic attribute information is for the sake of the system to sequence the printing task array and trace and inquires the original information in the subsequent customization procedure.

$$t_{basic-attribute} = \left(t_{user\ information}, t_{task\ description}, t_{task\ ID}, t_{timestamp} \right) \qquad (8.2)$$

Print attribute model is the regular description template established according to the distribution characteristics of 3D printing services and the physical and electrical attributes of 3D printers under the cloud manufacturing environment. For the precise matching of printing tasks with 3D printers, the model is qualitatively described only according to the physical and electrical attributes of 3D printers in this section. (For the quantitative description of the model, refer to the subsequent sections). Printing attribute information includes the following parameters: types of printers, minimum printing accuracy of printers, types of printable materials, colors of printing materials, model size, document format, and printing time. In terms of operating, in order to select the tasks of customized products, the user first of all upload the standardized document of 3D printing model such as STL, OBJ, etc., or select the appropriate 3D printing model document from the platform model library. According to the model document, the system calculates automatically and displays to the user the parameters of model size and printing time. Then, according to the system operating page, the user selects, confirms and submits such parameters as the type of printer, minimum printing accuracy of the printer, type of printable material, and color of printing material, as formula (8.3).

$$t_{print-attribute} = \left(t_{type}, t_{accuracy}, t_{space}, t_{printing-attribute}, t_{material}\right) \qquad (8.3)$$

To sum up, the printing task model provides a set of relatively perfect operating templates for the users to generate printing task orders in the 3D printing cloud platform. The system program of platform backstage can transform the operating information of users into regular printing task description information. From the angle of simulation, the printing task model provides the practically feasible solution for automatic generation of printing task description information. On the one hand, the printing task model classifies the relevant parameters of printing tasks finely and defines the normative data format framework for the sake of simulating each parameter according to their characteristics by the simulation method. On the other hand, as an operating template, the model can collect the actual operating information of users and can optimize the parameter generation algorithm by comparing the simulation data with the actual data so that the generated parameters are more approximate to the real data. Furthermore, by undertaking a considerable amount of simulation experiments with Monte Carlos method, it may be considered that the generated simulation data can be more really approximate to the real behavior of the user's actual operating.

2. Multi-source data integration model of printing services

While modeling for printing services, we not only take into account the static characteristics of 3D printers and analyze quantitatively all the parameter indicators of static characteristics, but also consider the dynamic characteristics of 3D printers to establish a universal dynamic data acquisition system, which can be used for real-time monitoring and automatic diagnosis of the health status of 3D printers. Therefore, printing services models can be classified as the static model of printing services and the dynamic model of printing services. Printing service model is as formula (8.4).

$$R = \left(r_{type}, r_{accuracy}, r_{space}, r_{file}, r_{material-type}, r_{material-color}, r_{status}, r_{fault-rate}, r_{fault-type}\right)$$
$$(8.4)$$

Static model of printing services

A 3D printer has a large number of performance features according to its working principle, size, component accuracy, printing process, design and assembly level and other aspects. When matching or scheduling 3D printers and printing tasks, it is necessary to select appropriate and valuable indicators to describe the features. The selected indicators may be different according to the needs of printing tasks and different matching or scheduling algorithms. Without losing generality, we describe the static features of printing service from six dimensions, they are the type of printers, printing accuracy, printable space, supportable document types, types of materials, and colors of materials according to the static attributes of existing universal 3D printers in the market.

Types of printers

The universal 3D printers currently include the following five types: direct metal laser sintering (DMLS) printers, selective laser sintering (SLS) printers, stereolithography apparatus (SLA) printers, digital light processing (DLP) printers, and fused deposition modeling (FDM) printers. Therefore, printer type models can be expressed as the set with the printer category, as formula (8.5):

$$r_{type} = \{DMLS, SLS, SLA, DLP, FDM\} \tag{8.5}$$

Print accuracy

According to the difference of the processing mechanism of 3D printers, printing accuracy has different meanings. For instances, in terms of FDM printers, since printing materials are generally printed with a nozzle, the diameter of the nozzle indicates the processing accuracy of such printers. In terms of SLS and SLA printers, since printing materials are generally modeled with the laser, the diameter of the laser transmitter indicates the processing accuracy of such printers. Therefore, the accuracy of 3D printers is subject to the sizes of printer extruders; accuracy can also be understood as the minimum layer size of printable models of 3D printers. Theoretically, the higher accuracy of one 3D printer extruder, the lower the value of its tape printing accuracy is, and thus the finer the model surface printed is. The currently common printers can generally include eight accuracy grades: 0.025, 0.05, 0.1, 0.2, 0.25, 0.4, 0.6, and 0.8 mm. The accurate model can be expressed as (8.6):

$$r_{accuarcy} = \{0.025 \text{ mm}, 0.05 \text{ mm}, 0.1 \text{ mm}, 0.2 \text{ mm}, 0.25 \text{ mm}, 0.4 \text{ mm}, \tag{8.6}$$
$$0.6 \text{ mm}, 0.8 \text{ mm}\}$$

Printable space

The printable space of 3D printers is generally defined depending on the maximum size of the printable model. By establishing the printing space specification indicators of 3D printing, we can obtain the printable space's quantization value $r_{space}p_{i-space}$. In order to ensure the printing space of small parts is an integer, in the study, we use the cubic meter as the minimum unit of printing space.

Supportable document types

The common CAD software such as AutoCAD, Solid Works, UG, and Pro/Engineer can be all used to design 3D printing models. The 3D-printer-readable document formats generated with such CAD software include STL, OBJ, AMF, and 3MF. The model of supportable document types can be expressed as in (8.7):

$$r_{file} = \{STL, OBJ, AMF, 3MF\} \tag{8.7}$$

Types of materials

Types of materials refer to the types of 3D printer consumables. Currently, the common consumables for 3D printing include eight types:

ABS, PLA, PET, TPU, UVCR, FUVCR, PA, METALALLOY, with the model as shown in (8.8):

$$r_{material-type} = \{ABS, PLA, PET, TPU, UVCR, FUVCR, PA, METALALLOY\} \tag{8.8}$$

Colors of materials

Colors of materials refer to the colors of printer consumables. The colors of materials are generally the set of a single color, including six colors: BLACK, WHITE, GRAY, HYALINE, SLIVER, and COLORFUL. The color of META-LALLOY material is generally referred to as SLIVER, while some rare colors and mixed colors formed with the multi-extruder printer are referred to as COLORFUL. The model is as shown in (8.9):

$$r_{material-color} = \{BLACK, WHITE, GRAY, HYALINE, SLIVER, COLORFUL\} \tag{8.9}$$

(Note: since some new dual-extruder or multi-extruder 3D printers have been developed and put into application in recent 2 years, the subsequent studies shall take into account the matching mode of multiple colors and mixed colors)

Dynamic model of printing services

In this study, we developed a set of model description methods for representing dynamic models of 3D printers. With 3D printer dynamic data acquisition system, it is possible to capture the data characterizing the dynamic attributes of 3D printers, so that the simulation process of supply-demand matching is more approximate to the real circumstances.

Fig. 8.2 presents the dynamic data acquisition system of 3D printers. On the embedded data collection board, Arduino, through the temperature and humidity sensor, vibration sensor, 3-axis accelerometer sensor, current sensor, and voltage sensor, the system collects the environmental and physical parameters of 3D printers and sends parameters to the dynamic attribute analyzer through I2C-based real-time clock module and Bluetooth 4.0 module. By deploying the pattern recognition algorithm, the dynamic attribute analyzer can analyze

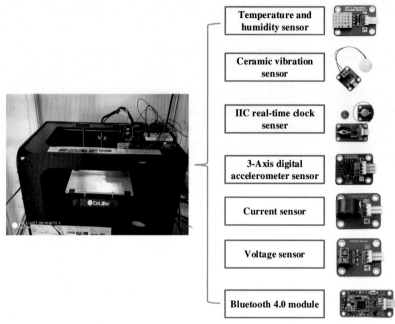

FIG. 8.2 The dynamic data acquisition system.

quickly and recognize the faults in the process of printing (e.g., starving, clogging, edge curl, and motor damage).

According to the foregoing description of the dynamic data acquisition system, the models characterizing the dynamic attributes of printing services cover the following three dimensions: working mode, fault rate, and fault type.

Work mode

Work mode indicates the in-use mode of 3D printers, including such five modes as ON, OFF, RUNNING, REPAIRING, and REPAIRED. When one 3D printer is observed with a fault, the mode of REPAIRING will be displayed. When the printer is fixed, its mode is displayed as REPAIRED. When the printer is already executing the printing task, the mode of RUNNING will be displayed. The working mode model is described as (8.10):

$$r_{status} = \{ON, OFF, RUNNING, REPAIRING, REPAIRED\} \qquad (8.10)$$

Fault rate

Fault rate means the frequency of faults observed with one printer access into the platform for use. In the model of (8.11), $F_{times}(i)$ represents the fault times of the printer and a serial number of printing times. The serial numbers of printing times can display visually the record of continual faults with the printer. $T_{times}(n)$ represents the total printing times of the printers.

$$r_{fault-rate} = \frac{F_{times}(i)}{T_{times}(n)} \tag{8.11}$$

Fault type

3D printer dynamic data acquisition system (see Fig. 8.2) is used for the real-time monitoring the fault of the 3D printer. In combination with fault tree and pattern recognition algorithm, the fault detection module is embedded in this system. When the system detects a printer failure, The system will push relevant dynamic information to the 3D printing cloud platform in real time. Faults with printers include four modes: STARVING, CLOGGING, EDGE_CURL, and MOTOR_DAMAGE. For the fault type model, see Formula (8.12). STARV-ING represents that the printer lacks printing materials, CLOGGING represents that the printer extruder is clogged, EDGE-CURL represents the edge curling of materials due to the temperature setting of the printer's working platform, and MOTOR-DAMAGE represents that the printer's motor is damaged.

$$r_{fault-type} = \{STARVING, CLOGGING, EDGE\_CURL, MOTOR\_DAMAGE\} \tag{8.12}$$

8.2.2 Capability indicator model

With the models of printing tasks and printing services established, in order to resolve the problem of supply-demand matching between printing tasks and printing services. In the study, we adopt the matching framework of a bipartite graph based on graph theory to support the matching of the two types of models. We extract and integrate the parameters of the original models (note: it is not mean here to deny the parameter setting of the original models. Since the parameters of the original models are set up for communication of model data in the network environment, it is necessary to adjust the models to adapt to the parameter-based expression form of bipartite-graph-oriented matching theory). Based on this, a set of the standardized matching process is established and detailed matching rules are formulated.

1. **Mathematic model of bipartite graph**
   Bipartite graph is a special model, in the graph theory, used to express the corresponding relation of two independent sets and is much suitable for describing the supply-demand matching relation between printing tasks and printing services herein. The mathematic expression formula is $G = <V, E>$, where V is the two independent vertex sets $V = V_1 \cup V_2$, $V_1 \cap V_2 = \Phi$ and E is the edge set for connection to any two vertexes. Therefore, it is necessary to establish the descriptive model of the vertex set and

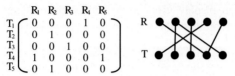

$$
\begin{array}{c}
\begin{array}{ccccc} R_1 & R_2 & R_3 & R_4 & R_5 \end{array}\\
\begin{array}{c} T_1\\ T_2\\ T_3\\ T_4\\ T_5 \end{array}
\left(\begin{array}{ccccc}
0 & 0 & 0 & 1 & 0\\
0 & 1 & 0 & 0 & 0\\
0 & 0 & 1 & 0 & 0\\
1 & 0 & 0 & 0 & 1\\
0 & 1 & 0 & 0 & 0
\end{array}\right)
\end{array}
$$

FIG. 8.3 Sample matrix description of non-weighted bipartite matching results.

edge set for bipartite graph matching. In the study, two vertex sets classification is undertaken for printing tasks and printing services in the description mode of $\{T, R\}$, where $R = \{R_1, R_2, R_3, \cdots, R_i, \cdots, R_m\}$ represents the vertex set of printing services and $T = \{T_1, T_2, T_3, \cdots, T_k, \cdots, T_n\}$ represents the vertex set of printing tasks. $\{E\}$ is the set of edges expressed in the form of matching degree.

The mathematic model of the bipartite graph can also be expressed more visually in the form of a conversion matrix. The bipartite graphs can include such two forms as weighted and non-weighted. Weight is the numerical form of edges. When weighted, the edge set $\{E\}$ is the real number set. When non-weighted, the edge set $\{E\}$ is the Boolean set of 0 or 1. Fig. 8.3 describes the matrix expression form of the non-weighted bipartite graph. $\{T\}$ and $\{R\}$ are used to correspond to the row and column of matrix respectively. If edge conforms to the matching condition, write 1 in the relevant matrix location. If not conforming, write in 0. In this way, the relevant transformation matrix is obtained. The figure is the matrix description of the random matching results based on 5 printing tasks and 5 printing services.

2. Capability indicator model based on bipartite graph

(1) Vertex set model

We establish one mathematic model of quantified indicators to realize the matching based on the bipartite graph, which can reflect the overall characteristics as well as can reflect all the dimensional attributes of printing tasks and printing services. According to the multi-source data integration model, the matchable parameters of two vertex sets include the following five dimensions: types of printer, color of materials, type of materials, printing accuracy, and printing space.

To resolve the matching calculation problem resulting from the large-scale printing orders in the cloud manufacturing environment, since the K-M algorithm complexity of vertex set ergodic matching based bipartite graph is $O(N^3)O(N^3)$, if all the tasks and services are matched directly, the calculation time will become an exponential growth. However, with printing services grouped according to the printer type model, the after-grouping algorithm complexity is $n_{\text{group-type}} \cdot \dfrac{O(N^3)}{O(n_{\text{group-type}})}$. Grouping can reduce greatly calculation consumption resulting from the large-scale matching. Therefore, we are grouping the types of printers.

(2) Edge set model

After printer grouping, we also necessary to build a matching algorithm according to the elements of other dimensions, to generate the edge set of a bipartite graph with weight. The algorithm of each element matching degree and total matching degree is as follows:

Matching degree of material color

Material color is one of the essential parameters for calculating the matching degree. The material color shall be matched subject to the type of printers. For instance, printing tasks with SILVER attribute representing universally metal alloy can be completed only by SLS and DMLS printers with the metal printing capability. Therefore, upon grouping the printer types primarily matched, the system will allocate the task set with SILVER attribute tag in material color E_{color} in the printing task array to SLS and DMLS printers with metal printing capability for matching.

In this study, consideration is given only to the requirement of the printers with unicolor processing capability for the color matching degree. The color set is mainly the common printing material colors. Thus, the color set is described as (8.13):

$$E_{color} = \{BLACK, WHITE, GRAY, HYALINE, SLIVER, COLORFUL\} \tag{8.13}$$

With consideration to universality, the consumable color of 3D printers $r_{i-color}$ is the element in E_{color}, and the color of printing tasks $t_{k-color}$ is the sub-set of E_{color}. The formula for calculating the color matching degree is as (8.14) follows:

$$g_{i-color} = \begin{cases} 1 & t_{k-color} \in r_{i-color} \\ 0 & t_{k-color} \notin r_{i-color} \end{cases} \tag{8.14}$$

Since the consumables of 3D printers are generally provided with the set of different colors, the matching degree of colors is 0 only when the supply of material colors is inadequate. Therefore, the weight of color matching degree is 1 or 0. (Note: the parameter α in parameter α in $t_{k-color}(\alpha)$ represents the color not matched, which can be used in notifying the supplement of consumable. See the detailed interpretation of the case study.)

Matching degree of printing accuracy

The accuracy set of printers is described as (8.15) follows:

$$E_{accuracy} = \{0.025\ mm, 0.05\ mm, 0.1\ mm, 0.2\ mm, 0.25\ mm, 0.4\ mm,$$

$$0.6\ mm, 0.8\ mm\} \tag{8.15}$$

The accuracy model of printing tasks is $t_{k-accuracy}(\delta)$, and the accuracy model of printing services is $r_{i-accuracy}$. In terms of processing mechanism of 3D printers, the higher the accuracy of one 3D printer nozzle, the lower the value of its printing accuracy. Therefore, in the matching process, printing tasks can only be matched to the printing services with lower accuracy value. Additionally, since the printing services with high printing accuracy are used to process the printing tasks with excessively low accuracy, it will not only cause a waste of service but also may easily cause clogging of printing nozzle, in the formula, we set δ as the span value of accuracy and restrict the accuracy of printing services can be at most two grades lower than the accuracy of printing tasks, for the grade gap of accuracy will also affect the weight of matching degree. The calculation formula is as (8.16):

$$G_{i-accuracy} = \begin{cases} 1 & \delta = 0 & t_{(k-accuracy)} \geq r_{(i-accuracy)} \\ 0.85 & \delta = 1 & t_{(k-accuracy)} \geq r_{(i-accuracy)} \\ 0.7 & \delta = 2 & t_{(k-accuracy)} \geq r_{(i-accuracy)} \\ 0 & & t_{(k-accuracy)} < r_{(i-accuracy)} \end{cases} \tag{8.16}$$

Matching degree of printable space

Matching degree of printable space means the proportion of 3D model volume $t_{k-space}$ and printer's actual printable space $r_{i-space}$. $r_{i-space}$ is a fixed value obtained according to the particular model of printer. In order to assure that the model can be in priority matched to the printer with the approximate volume, the printing task model $t_{k-space}$ can only be matched to the printer with a bigger volume, the weight of spatial matching degree is set as the ratio of tasks and printer, with the calculation formula (8.17):

$$g_{i-space} = \begin{cases} \dfrac{t_{k-space}}{r_{i-space}} & t_{k-space} \leq r_{i-space} \\ 0 & t_{k-space} > r_{i-space} \end{cases} \tag{8.17}$$

Matching degree of material type

Since the set of material types is the standard consumables types of 3D printers, the set of materials is described as (18) follows:

$$E_{material} = \{ABS, PLA, PET, TPU, UVCR, FUVCR, PA, METALALLOY\} \tag{8.18}$$

In the study, 3D printers of a single material are taken mainly as the study objects (note: new printers with dual-nozzle or multi-nozzle for printing different materials at the same time are already currently marketed, but such printers are not included in the category of this study). The material type of 3D printers is $r_{i-material}$, the material demand of printing tasks is $t_{k-material}$. The formula for calculating the matching degree of material types is as (8.19) follows:

$$g_{i-material} = \begin{cases} 1 & t_{k-material}(\beta) \in r_{i-material} \\ 0 & t_{k-material}(\beta) \notin r_{i-material} \end{cases} \tag{8.19}$$

Since 3D printer are generally provided with the set of different consumables, the matching degree for types of materials will generally be 0 only when the supply of material type is inadequate. Therefore, the matching degree of material types is 1 or 0. (Note: the parameter β in $t_{k-material}(\beta)$ represents the material type not matched to be used for the notice of consumables supply. For the details, see the interpretation of the matching rules in the case study.)

Total matching degree

The total matching degree is the product of the preceding matching degrees, with the calculation formula as (8.20):

$$G_i = g_{i-color} \cdot g_{i-accuracy} \cdot g_{i-space} \cdot g_{i-material} \tag{8.20}$$

When the total matching degree is calculated, it is the weight of edge in the bipartite graph. The final result of G_i is a decimal number not more than 1. When the value is more approximate to 1, it means the matching degree is higher. According to the total matching formula, as long as the matching degree of any one of the dimensions does not meet the requirements, its value is 0. In the matching process base on the bipartite graph, we define the non-zero value of the total matching degree to generate a solid edge, and the value of zero to generate a dashed edge.

8.2.3 Model-based matching process and matching rules

We adopt the bipartite graph based on graph theory to support the matching of models. In the process of model instantiation, we first construct two vertex sets {T} and {R} of the bipartite graph according to the multi-source data integration model of printing tasks and printing services respectively. By instantiating the static and dynamic models of the printing services, the static matching results and the dynamic matching results will be formed respectively. Then, we calculate the edge set {e} of the bipartite graph through the total matching degree of the capability indicator model. In this way, the vertexes set and edge set of the bipartite graph are generated by a model-driven method. In the process of model matching, we use three typical matching algorithms to compare the matching rate of instantiated models. The three algorithms include the Kuhn-Munkres algorithm (K-M algorithm), Kruskal improved algorithm (Kruskal+ algorithm), Prime improved algorithm (Prime+ algorithm). Among them, K-M algorithm is a weighted binary matching algorithm based on the augmented path, the purpose of this algorithm is to find the maximum matching rate of two vertex sets of the bipartite graph. Kruskal+ algorithm and Prime+ algorithm are the maximum edge weight matching algorithm based on minimum spanning tree. The goal of Kruskal improved algorithm is to ensure the maximum matched service value, while the goal of Prime improved algorithm is based on the priority of printing task timestamp. From an algorithm implementation perspective, K-M algorithm belongs to the Hungarian algorithm based on Depth-First-Search (DFS) strategy, whose algorithm complexity is $O(N3)$. Kruskal+ and Prime+ algorithms belong to the greedy algorithm based on Breadth-First-Search (BFS) strategy, whose algorithm complexity is respectively $O(E \cdot logV)$ and $O(V2)$.

As explained by the above process, the cloud 3D printing matching method based on dynamic and static data has the following highlights: (1) We construct a normal cloud 3D printing model, and can fetch the real data of the printing task and the printing service model. In this way, in the simulation process, we can generate a large number of simulation data according to the data samples. (2) We divide the model of printing services into the dynamic model and static model. By analyzing the experimental results of static matching and dynamic matching, we can objectively summarize the matching effects of different types of printers on different matching algorithms.

1. **Matching process**

    The matching process is given as follows:
    - **(1)** Load the printing task queue, which is sequenced according to the element of $t_{timestamp}$ in the vertex set of printing tasks;
    - **(2)** According to the type of printers in the printing tasks, load the same printing services matched, and generate the bipartite graph;
    - **(3)** Load the static parameters of the printing tasks queue and printing services queue;
    - **(4)** Execute the total matching degree algorithm and generate the weighted edge;
    - **(5)** Cycle step 1 to step 5 50 times and respectively run three matching algorithms;
    - **(6)** Generate the bipartite graph and transformation matrix of the current matching results;
    - **(7)** Generate static matching result;
    - **(8)** Load the dynamic parameters of the printing tasks queue and printing services queue;
    - **(9)** Cycle step 1to step 9 50 times and respectively run three matching algorithms;
    - **(10)** Generate the bipartite graph and transformation matrix of the current matching results;
    - **(11)** Generate dynamic matching result.

2. **Matching rules**

    In order to assure the matching process runs correctly in the distributed manufacturing environment, it is necessary to formulate the relevant matching rules for the smooth execution of each step in the matching procedure. We establish the static and dynamic matching rules respectively according to the static and dynamic parameter characteristics of printing tasks and printing services.

Static matching rules

(1) According to Hall theorem [8], The number of printing task queue must be less than or equal to the number of available printing services. Where $R = \{R_1, R_2, R_3, \cdots, R_i, \cdots, R_m\}$ represents the vertex set of printing services and $T = \{T_1, T_2, T_3, \cdots, T_k, \cdots, T_n\}$ represents the vertex set of printing tasks. So, $n \leq m$.

(2) The static model is the foundation for matching. Static model parameters need to be set according to the actual characteristics of printing tasks and printing services. For example, the DMLS 3D printer belongs to the metal printer, the material type of which is METALLALLOY and the material color is SLIVER. In addition, from the point of view of printing accuracy, the accuracy range of all printers is

$$E_{accuracy} = \{0.025 \text{ mm}, 0.05 \text{ mm}, 0.1 \text{ mm}, 0.2 \text{ mm}, 0.25 \text{ mm}, 0.4 \text{ mm},$$

$$0.6 \text{ mm}, 0.8 \text{ mm}\}$$

However, different types of 3D printers have different printing accuracy ranges. In general,

$$R_{accuarcy} - FDM > R_{accuarcy} - DLP > R_{accuarcy} - SLA > R_{accuarcy} - SLS >$$

$$R_{accuarcy} - DMLS$$

Therefore, building a set of matching parameter rules is an important prerequisite for accurate matching of supply and demand models.

(3) When the parameters of a printing task are set to very stringent conditions, the printing task may not match the appropriate printing services. As for the printing task failing to match with printing services for many times, it is necessary to re-define the task demand. The threshold value means the times that printing task failing to match the printing service. When the threshold value is equal to 3, this printing task need user to re-set parameters.

Dynamic matching rules

In the dynamic matching rules, since matching results and transformation matrix based on the bipartite graph are already generated, this study adds the dynamic matching optimization stage innovatively based on the dynamic data acquisition system in the matching simulation procedure. Dynamic matching optimization provides the re-adjustment for the matching results with the dynamic data acquisition system according to the dynamic model of 3D printers (work mode, fault rate, and fault type) and optimizes the matching results of the static matching process using the approach of virtual-actual combination.

As discussed above, the dynamic matching rules are described as follows:

(1) In the matching process, printing services are added with three new characteristics with dynamic attributes (working mode, fault rate, and fault type);

(2) Printing services in the mode of $R_{status} = $ REPAIRING will be defined as a concealing mode not for the matching process until restoring to the mode of $R_{status} = $ REPAIRED;

(3) In the process of matching, allocate printing tasks in priority with printing services of a lower fault rate, i.e., fault rate will affect the edge weights in the bipartite graph;

(4) When the fault type is in the category of repairable fault ($r_{fault-type} = $ {STARVING, CLOGGING, EDGE_CURL}), enter the reprint process and update the fault rate of the device. When the fault type is in the category of unrepairable fault ($r_{fault-type} = $ MOTOR_DAMAGE), this printing task will be temporarily removed from the printing queue, the dynamic attribute of the printing service is set as the mode of $r_{fault-status} = $ REPAIRING until the fault is repaired;

(5) When $r_{fault-type} = $ STARVING information appears the system will automatically feedback $t_{k-color}(\alpha)$ and $t_{k-material}(\beta)$ information to the printing service management center, reminding the maintenance staff for prompt supplementing the printing consumables;

(6) 3D Printer with continual faults for several times should be considered for removing the 3D printing platform($F_{times}(i)$ represents the fault times of the printer and a serial number of printing times, when this sequence $F_{times}(i)$ displays fault for 5 consecutive times, the 3D printer will be logged off from the 3D printing platform).

8.3   Scheduling of distributed 3D printing services

3D printing service scheduling consists of two processes including service matching and task scheduling. Through matching 3D printing tasks to services, candidate services of each task can be obtained from service pool in the cloud platform based on the matching method in the last section. But there are still two issues need to be solved in order to generate optimal service scheduling solutions. On one hand, we need to exactly match a most suitable service for each task in the service matching process because there may be multiple candidate services for a task. On the other hand, we need to decide when to start executing each task separately on a matched service in the task scheduling if this service is matched by multiple tasks in the service matching process. During 3D printing service scheduling, a service matching solution and a task scheduling solution make up a service scheduling solution, which can be transformed into a intelligent optimization problem. A 3D printing service scheduling method (3DPSS)

[9] from aspects of optimization objectives, constraints and optimization algorithm is introduced.

8.3.1 Optimization objective

Fig. 8.4 shows a result of distributed 3D printing service matching. We can see that there are multiple distributed 3D printing services 3DPS-1, 3DPS-2, 3DPS-3, 3DPS-4 and multiple 3DP tasks 3DPT-1, 3DPT-2, 3DPT-3 on the map. Task 3DPT-1 is matched to 3DPS-1, 3DPS-3 and 3DPS-4. Task 3DPT-2 is matched to 3DPS-2 and 3DPS-3. Task 3DPT-3 is matched to 3DPS-2, 3DPS-3 and 3DPS-5. Only one service will be matched for each 3D print task from its all candidate services. For example, tasks 3DPT-1, 3DPT-2 and 3DPT-3 matches 3DPS-1, 3DPS-3 and 3DPS-5. The service matching process is combined with task scheduling to generate service scheduling solutions.

Manufacturing services in the CMfg platform are ultimately provided to service consumers. A consumer may submit multiple 3D printing tasks to the CMfg platform at the same time. Each 3D printing task includes a specified 3D printing model and individualized requirements of the consumer for this task such as printing material, printing accuracy and task cost.

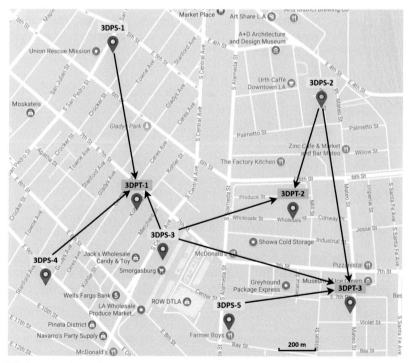

FIG. 8.4 Matching of distributed 3D printing services.

The advantages of 3D printing technology mainly include rapid prototyping and personalized customization. For CMfg 3D printing services, it is a crucial target to quickly complete the 3D printing tasks of customers and timely deliver these tasks to the customers. Both the unit printing cost and printing preciseness of a specific 3D printer are generally determinate. But the task delivery time is a more complex variable than cost and preciseness. There are more room for the time attribute to be optimized than cost and preciseness in the scheduling process. Therefore, the task cost and printing preciseness are both considered as constraints of the scheduling problem while the task completion time is considered as the optimization objective to embody the rapidity of 3D printing. For a specific service demander Di, he wants his multiple 3D printing tasks to be completed as soon as possible in the CMfg platform. But for the CMfg platform, task delivery time of different service demanders should be considered together in the optimization objective function of 3D printing service scheduling to reduce the total task delivery time of all service demanders. Therefore, we regard the average latest task delivery time of all service demanders as the optimization objective of 3D printing service scheduling problem.

Besides, we take into account that different service demanders have different priorities in the optimization objective function. A service demander's priority is represented by a weight coefficient of this demander's latest task delivery time in the scheduling optimization objective function. The priority of customer Di is represented by r_i. The mathematical expression of the optimization objective is given by (8.21).

$$\min \frac{\sum_{i=1}^{n} r_i F_i}{\sum_{i=1}^{n} r_i} \tag{8.21}$$

The latest product delivery time F_i of service demander D_i is equal to the delivery time of the task lastly delivered to D_i through logistics. F_i is calculated by (8.22).

$$F_i = \max_{j=1}^{d_i} b_{i,j} \tag{8.22}$$

The delivery time of j-th task T_{ij} of service demander D_i is equal to the sum of start time of T_{ij}, printing time of T_{ij} on the matched service and logistics time of T_{ij} from the matched service provider S_k to D_i, as shown by (8.23).

$$b_{i,j} = a_{i,j} + f_{i,j} + l_{i,k_{i,j}} \tag{8.23}$$

Printing time of 3D printing tasks consists of four parts which are warm up time, processing time, cooling time and takedown time, and processing time is the main part of printing time. Processing time of a 3D printing task is related to the size of the 3D model and operating parameters of the matched 3D printing services. Although the warm up time, cooling time and takedown time of different 3D printing services are generally of different values, these three time attributes of a specific 3D printing service are always constants for different tasks. The printing time of task T_{ij} on service S_k can be calculated by (8.24). Logistics time is related to logistics speed and geographical distances between service demanders and matched service providers as shown in (8.25). Based on (8.22)–(8.25), the optimization objective of 3D printing service scheduling problem in CMfg can be obtained and the mathematical expression is given by (8.26).

$$f_{i,j} = \frac{g_{i,j}}{h_{k_{i,j}}} + H_{k_{i,j}} + Q_{k_{i,j}} + R_{k_{i,j}} \tag{8.24}$$

$$l_{i,k_{i,j}} = \frac{\sqrt{\left(x_i - X_{k_{i,j}}\right)^2 + \left(y_i - Y_{k_{i,j}}\right)^2}}{\alpha} \tag{8.25}$$

$$\min \frac{1}{\frac{1}{n}\sum_{i=1}^{n} r_i} \sum_{i=1}^{n} r_i \max_{j=1}^{d_i} \left(a_{i,j} + \frac{g_{i,j}}{h_{k_{i,j}}} + H_{k_{i,j}} + Q_{k_{i,j}} + R_{k_{i,j}} \right.$$

$$\left. + \frac{\sqrt{\left(x_i - X_{k_{i,j}}\right)^2 + \left(y_i - Y_{k_{i,j}}\right)^2}}{\alpha} \right) \tag{8.26}$$

For i and j ($1 \leq i \leq n$, $1 \leq j \leq di$), it is needed to find an optimal 3D printing service S_k and an optimal start time aij for task T_{ij} to minimize the value of objective function.

8.3.2 Constraints

1. Constraints of matched services

The service ID k_{ij} of all matched service constitute a Z-dimensional service matching vector η as given in (8.27).

$$\eta = (k_{1,1}, \cdots, k_{1,d_1}, k_{2,1}, \cdots, k_{2,d_2}, \cdots, k_{n,d_n}) \tag{8.27}$$

Constraints of matched services in 3D printing service scheduling are actually the matching rules of different service attributes in 3D printing service matching process because the objective of 3D printing service matching is to generate a candidate service set for each 3D printing task. These matching rules of different service attributes constitute an inequality group as given by (8.28) and (8.29).

$$\begin{cases} \min\left(u_{i,j},\ v_{i,j}\right) \leq \min\left(U_{k_{i,j}},\ V_{k_{i,j}}\right) \\ \max\left(u_{i,j},\ v_{i,j}\right) \leq \max\left(U_{k_{i,j}},\ V_{k_{i,j}}\right) \\ w_{i,j} \leq W_{k_{i,j}} \\ m_{i,j} = M_{k_{i,j}} \\ P_{k_{i,j}} \leq p_{i,j} \\ q_{i,j} \leq \dfrac{g_{i,j}c_i}{h_{k_{i,j}}P_{k_{i,j}}} + \beta u_{i,j}v_{i,j}w_{i,j}\sqrt{\left(x_i - X_{k_{i,j}}\right)^2 + \left(y_i - Y_{k_{i,j}}\right)^2} \end{cases} \qquad (8.28)$$

where

$$\begin{cases} 1 \leq i \leq n \\ 1 \leq j \leq d_i \end{cases} \qquad (8.29)$$

2. Constraints of start time of tasks

Task priority vectors are applied to deal with the constraints of start time a_{ij} of tasks. The task priority vector δ is a Z-dimensional vector as defined in (8.30). Each element in vector δ is a real number and all elements are unequal to each other. Element λ_i represents the priority of the i-th task. Note that the order of these elements in the task priority vector is independent of the values of elements.

$$\delta = (\lambda_1,\ \lambda_2,\ \cdots,\ \lambda_Z) \qquad (8.30)$$

Based on the task priority vector, the task scheduling strategy is presented. If multiple tasks match the same 3D printing service, then tasks with higher priority values are to be executed earlier on the matched service than those tasks with lower priority values. Normally, a 3D printing service can only execute one 3D printing task at the same time. Thus, a service matching vector η and a task priority vector δ uniquely determine a solution of CMfg 3D printing service scheduling problem.

8.3.3   Optimization algorithm

The genetic algorithm is applied to generate optimal 3D printing service scheduling solutions for all tasks. An improved GA algorithm is applied in 3DPSS method to solve the 3D printing service scheduling problem in CMfg as shown in the following example. Suppose there are three service providers D_1, D_2, D_3 and five available 3D printing services S_1, S_2, S_3, S_4, S_5, as shown in Fig. 8.5. Service demanders D_1, D_2 and D_3 submit two, three and three tasks, respectively. The connected relations from tasks to services in Fig. 8.5 illustrate the service matching results in which candidate services of each task can be obtained. From the perspective of 3D printing services, matching relationship between 3D printing tasks and services reflects all matched tasks of each 3D printing service, as shown in Fig. 8.6.

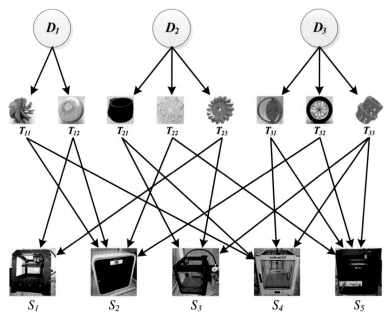

FIG. 8.5 Matching results of distributed 3D printing services for demanders' tasks.

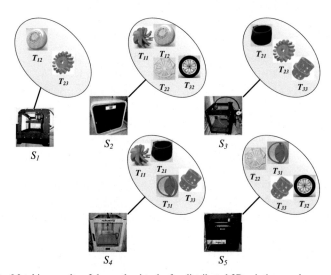

FIG. 8.6 Matching results of demanders' tasks for distributed 3D printing service.

Based on the service matching results in Fig. 8.5, service matching vector η can be applied to obtain the matched services of all tasks. Assume that the service matching vector η is (2,1,4,2,3,5,5,4). Fig. 8.7A shows the service matching results. We see that services S_1 and S_3 are both matched by one task while services S_2, S_4, and S_5 are matched by multiple tasks.

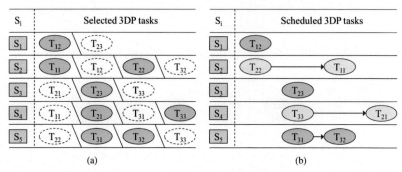

FIG. 8.7   Matching and scheduling of distributed 3D printing services.

TABLE 8.1 Chromosome coding in the GA.

$T_{i,j}$	T_{11}	T_{12}	T_{21}	T_{22}	T_{23}	T_{31}	T_{32}	T_{33}
Gene	1	2	3	4	5	6	7	8
η	2	1	4	2	3	5	5	4
δ	4	6	1	5	8	3	2	7

Task priority vector can then be applied to generate service scheduling results according to the service matching results in Fig. 8.7A. Assume that the task priority vector δ is (4,6,1,5,8,3,2,7). For tasks T_{22} and T_{11} on service S_2, T_{22} is executed earlier than T_{11} because the priority of T_{22} ($\lambda_4 = 5$) is higher than that of T_{11} ($\lambda_1 = 4$). In the same way, the processing sequences of all tasks on each service can be separately determined. Fig. 8.7B gives the generated task scheduling results of all services.

GA is applied to search the 3D printing service scheduling solution space and generate an optimal solution. In GA, we use a service matching vector η and a task priority vector δ to code service scheduling solutions. For a 3D printing service scheduling problem as given in Fig. 8.5, a chromosome example is shown in Table 8.1.

In chromosomes, value range of the i-th gene is the set of service ID matched to the i-th task in η. In this example, the first gene in η corresponds to task T_{11}. The set of service ID matched to T_{11} is {2, 4}. Therefore, the value range of the first gene in η of this chromosome is {2, 4}. In chromosomes, value ranges of genes in vector δ are {1, 2, ..., Z} and all genes are different from each other. The workflow of the proposed 3DPSS method is given by Fig. 8.8.

In the 3DPSS method, because a scheduling solution consists of a service matching vector and a task priority vector, the number of variables is twice

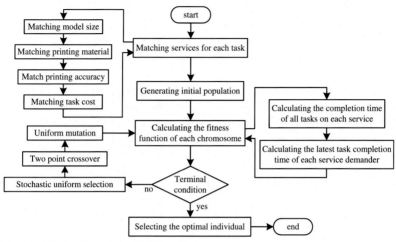

FIG. 8.8 The workflow of the 3DPSS method.

the size of the number of tasks. In the GA, the population size is 20. Stochastic uniform strategy is used as the matching function and the step size is 0.05. Two point strategy is used as the crossover function and the value range of the two random integers is [1, 2Z]. Besides, the uniform mutation method is applied and the mutation rate is set to 0.01. Compared with the other optimization algorithms, the 3DPSS method has the following characteristics:

- A task priority vector is proposed to describe the permutations of different tasks in the same service, which solves the complex multi-task sequencing problem in the same service.
- The task priority vector and service matching vector are combined to constitute a multi-task scheduling solution, which reduces the scheduling solution space and speeds up the solving process.
- A two point crossover strategy is applied to improve the searching capability of the 3DPSS method.

8.4 Conclusion

When a large number of orders are submitted to the 3D printing platform, one of an important capability of the cloud platform is to find the most suitable printing device for each printing task quickly and accurately. The first part of this chapter describes the printing task and printing service by comprehensively considering the needs of users, the characteristic parameters of the printer itself and the real-time state parameters of the printing process. On this basis, the capability indicator model and matching rules for supply–demand matching are given. Then, the printing task scheduling problem is further studied. Then, the exact matching and scheduling of printing tasks are further studied. When a task has multiple optional printing services, the most suitable one needs to be selected. While, when a printing service is matched by multiple tasks, it is necessary to determine the execution order and execution time of these tasks. To solve thses problems, a 3D printing service scheduling method is proposed based on the intelligent optimization method.

Supply and demand matching and task scheduling are core issues in cloud manufacturing service platforms, of course including cloud 3D printing platforms. The current research cannot meet the needs of the actual operations of the platforms. There are a lot of uncertainties in the requirements and services provided by the platform, such as user modifying orders, service failures, unstable service qualities, and so on. These uncertainties in the cloud platform are much more serious than the similar situation encountered by the traditional job shop scheduling problem. Because in the cloud platform, the transactions between the supplier and the demander are often carried out without establishing mutual trust through long-term cooperation's. In such a complex environment, how to quickly and efficiently realize dynamic matching and real-time scheduling, so as to ensure that the task can be completed on time and with high quality, is an important research direction in the future.

References

[1] L.F. Zhou, L. Zhang, L. Ren, Y.J. Laili, Matching and selection of distributed 3D printing services in cloud manufacturing, in: Proceedings of the 43rd Annual Conference of the IEEE Industrial Electronics Society. Beijing, China, 2017.

[2] A. Antonio, B. Mattia, C. Marco, Warpage of FDM parts: experimental tests and analytic model, Robot. Comput. Integr. Manuf. 50 (2018) 140–152.

[3] T. Yedige, S.H. Geok, L. Wen-feng, Nozzle condition monitoring in 3D printing, Robot. Comput. Integr. Manuf. 54 (2018) 45–55.

[4] T.C.T. Chen, Y.C. Lin, A three-dimensional-printing-based agile and ubiquitous additive manufacturing system, Robot. Comput. Integr. Manuf. 55 (2019) 88–95.

[5] C. Shi, L. Zhang, J.G. Mai, Z. Zhao, 3D printing process selection model based on triangular intuitionistic fuzzy numbers in cloud manufacturing, Int. J. Model. Simul. Sci. Comput. 8 (2) (2016) 1750028.

[6] Z. Zhao, L. Zhang, J. Cui, 3D print task packing algorithm based on rectangle packing in cloud manufacturing, in: Proceedings of Chinese Intelligent Systems Conference, Mu Dan Jiang, China, Springer, 2017, pp. 21–31.

[7] X. Luo, L. Zhang, L. Ren, Y.J. Laili, A dynamic and static data based matching method for cloud 3D printing, Robot. Comput. Integr. Manuf. 61 (2020), 101858.

[8] P. Hall, On representatives of subsets, J. Lond. Math. Soc. 10 (1935) 26–30.

[9] L.F. Zhou, L. Zhang, Y.J. Laili, C. Zhao, Y.Y. Xiao, Multi-task scheduling of distributed 3D printing services in cloud manufacturing, Int. J. Adv. Manuf. Technol. 96 (9–12) (2018) 3003–3017.

Chapter 9

3D printing process management

With the popularity of 3D printing platform, a large number of 3D printing orders will be placed on the platform at the same time. Then, how to improve the printing efficiency as much as possible through fine management and optimization has become an urgent problem to be solved. This chapter will discuss the management methods of printing process from two aspects, they are multi-task parallel printing and task packing.

9.1 Multi-task parallel printing for complex products

Orders on the 3D printing cloud platform may be simple products or complex products. For simple and small products, the design model can be produced through one-time printing, which only needs one 3D printer to be matched with. For a complex product that's composed of multiple parts with different sizes, shapes, materials and properties, it is necessary to decompose the product into multiple parts (Fig. 9.1). The processing parameters of different parts are different. Taking the precision as an example, if the parts with low precision requirements are processed with high-precision printers, then the printing costs will increase; if multiple parts are processed on one 3D printer, the printing time will increase. Therefore the most suitable printers need to be selected for different parts to form multiple parallel processing subtasks, and finally complete the production task of the whole product. Through the reasonable assignment and management of multiple printing subtasks, the overall production time and cost can be reduced to the greatest extent while meeting the product's requirements.

9.1.1 Problem definition

For a printing task T with n parallel subtasks, $T = (T_1, T_2, ..., T_n)$, match the δ (i) services that can meet the processing requirements from the cloud platform according to the processing parameters of ith subtask, to constitute a matching service set L_i.

Customized Production Through 3D Printing in Cloud Manufacturing
https://doi.org/10.1016/B978-0-12-823501-0.00004-3

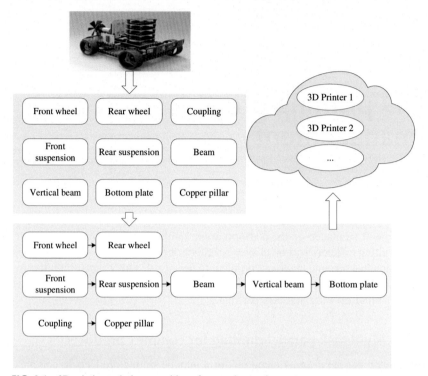

FIG. 9.1   3D printing task decomposition of a complex product.

Let Y be a finite set of all the elements of the above matching service sets, then $L = (L_1, L_2, ..., L_n)$ is a subset family of Y [1,2]. A family $e = (e_1, e_2, ..., e_n)$ of the elements of Y is called a candidate solution of the task T, if e satisfies $e_1 \in L_1, e_2 \in L_2, ..., e_n \in L_n$. If the elements $e_1, e_2, ..., e_n$ in a candidate solution are different, $(e_1, e_2, ..., e_n)$ is called a feasible solution of the task T.

Based on the above assumptions, the general form of parallel processing service composition optimization problem can be described as:

$$\begin{cases} \min g(e) \\ s.t.\ e_i \text{different}, e_i \in L_i, i = 1, 2, ..., n \end{cases} \quad (9.1)$$

Where min g (e) generally refers to the optimization objective. According to different user needs, different optimization objectives can be adopted, such as the optimization based on the shortest service time, the optimization based on the lowest total cost or the optimization based on the largest comprehensive index.

Let $g_T()$ is the evaluation function of service time, and the service time of parallel tasks is the maximum value of the service time of each subtask, then the optimization problem based on the shortest parallel service time can be expressed as:

$$\begin{cases} \min \ \max g_T(e_i) \\ s.t. \ e_i \text{different}, e_i \in L_i, i = 1, 2, ..., n \end{cases} \qquad (9.2)$$

Let $g_C()$ is the evaluation function of the service cost, and the service cost of the task is the sum of the service costs of each subtask, then the optimization problem based on the lowest total service cost can be expressed as:

$$\begin{cases} \min \ \sum_{i=1}^{n} g_C(e_i) \\ s.t. \ e_i \text{different}, e_i \in L_i, i = 1, 2, ..., n \end{cases} \qquad (9.3)$$

If the mean value of the service matching degree $g_M(e_i)$ of each subtask is used as the comprehensive optimization objective, the optimization problem can be expressed as:

$$\begin{cases} \max \ \sum_{i=1}^{n} \dfrac{g_M(e_i)}{n} \\ s.t. \ e_i \text{differen}, e_i \in L_i, i = 1, 2, ..., n \end{cases} \qquad (9.4)$$

9.1.2   Combinatorial optimization method of parallel processing services

For the parallel processing service combination optimization problem of Eqs. (9.2)–(9.4), the candidate service set of each subtask is solved in parallel, then select n different services from each subtask candidate service set to form a composite solution. When the number of subtasks and candidate services is large, there will be a combination explosion, which needs to be solved by optimization algorithm. The solving process is analyzed as follows by taking the comprehensive index optimization of formula (9.4) as an example.

1. Basic concepts

Set subtask J to have δ (j) applicable services. The service ordered set $S_j = (S_{1j}, S_{2j}, ..., S_{\delta(j)j})$ in descending order of matching degree meets

$$g_M(S_{1j}) \geq g_M(S_{2j}) \geq ... \geq g_M(S_{\delta(j)j}) \qquad (9.5)$$

Where $g_M()$ is the comprehensive matching degree of a service. The candidate solutions of subtasks can only be selected from the corresponding ordered set of services. After building the ordered set, the information contained in S_{ij} can be expressed as

$$< i, j, g_M(S_{ij}), N(S_{ij}) >$$

Where i represents the sequence number of subtasks, j represents the sequence number of services in the ordered set, and $g_M(S_{ij})$ is the matching degree value, and $N(S_{ij})$ represents the unique identification of the service in the cloud platform., which is expressed in numerical values. When the identifications of different services are not equal to each other, these services are different from each other. Obviously, in an ordered set of services, services are different from each other.

Set m to be the minimum number of elements in each ordered service set, i.e.

$$m = \min \delta(j) \; j = 1,...,n \qquad (9.6)$$

Based on (9.5) and (9.6), intercept the first m services in each ordered set of service as a column, a matrix with M rows and N columns can be constructed, which is called the preferential candidate service matrix.

$$\begin{pmatrix} S_{11} & \cdots & S_{1n} \\ \vdots & \ddots & \vdots \\ S_{m1} & \cdots & S_{mn} \end{pmatrix}$$

For this matrix, we can obtain two properties that contribute to the heuristic solution of the service composition optimization problem in (9.4):

Property 1: when $m \geq n$, there is at least one set of different services in the preferential candidate service matrix to form a feasible solution.

Proof: Firstly, the services in the same column in the service matrix are arranged in descending order of identification to form a new column; Then adjust the column order according to the size of the first value of each column, and finally construct the following matrix:

$$N = \begin{pmatrix} N_{11} & \cdots & N_{1n} \\ \vdots & \ddots & \vdots \\ N_{m1} & \cdots & N_{mn} \end{pmatrix} \qquad (9.7)$$

Where, elements of each column are arranged in descending order to satisfy

$$N_{1j} > N_{2j} > ,... > N_{mj} \; j = 1,...,n \qquad (9.8)$$

And the elements in the first row satisfy

$$N_{11} \geq N_{12} \geq ,... \geq N_{1n} \qquad (9.9)$$

By combining (9.8) with (9.9), we can obtain

$$N_{11} \geq N_{12} > N_{22}$$

$$\cdots$$

$$N_{11} \geq N_{1n} > N_{2n}$$

Hence

$$N_{11} > N_{22}$$

$$\cdots$$

$$N_{11} > N_{2n}$$

That is

$$N_{11} > \max N_{2j} \ j = 2,...,n \tag{9.10}$$

Take rows 2 to m and columns 2 to n of matrix N to form a matrix

$$\begin{pmatrix} N_{22} & \cdots & N_{2n} \\ \vdots & \ddots & \vdots \\ N_{m2} & \cdots & N_{mn} \end{pmatrix}$$

Adjust the column order to construct a new matrix

$$\mathbf{N}^1 = \begin{pmatrix} N_{22}^1 & \cdots & N_{2n}^1 \\ \vdots & \ddots & \vdots \\ N_{m2}^1 & \cdots & N_{mn}^1 \end{pmatrix}$$

Where, the elements in the first row satisfy

$$N_{22}^1 \geq N_{23}^1 \geq ,... \geq N_{2n}^1 \tag{9.11}$$

Because the order of the elements in the column is not adjusted, the elements in each column still meet the descending order.

$$N_{2j}^1 > N_{3j}^1 > ,... > N_{mj}^1 \ j = 2,...,n \tag{9.12}$$

By combining (9.11) with (9.12), we can obtain

$$N_{22}^1 > N_{33}^1$$

$$\cdots$$

$$N_{22}^1 > N_{3n}^1$$

That is

$$N_{22}^1 > \max N_{3j}^1 \ j = 3,...,n \tag{9.13}$$

According to the process of constructing N from matrix N_1, we can obtain

$$N_{22}^1 = \max N_{2j} \ j = 2,...,n \tag{9.14}$$

By combining (9.10) with (9.13) and (9.14), we can obtain

$$N_{11} > N_{22}^1 > \max N_{3j}^1 \tag{9.15}$$

Repeat (9.10)–(9.15), when $m \geq n$, other matrices $N^2, N^3, ..., N^{n-1}$ can be constructed successively, and the elements in the first row and first column of each matrix satisfy

$$N_{11} > N_{22}^1 > N_{33}^2 > ,..., > N_{nn}^{n-1} \tag{9.16}$$

The n elements are different services and they are in different columns, so they belong to the feasible solution.

Property 2: if elements in the first row of the preferential candidate service matrix form a set of feasible solutions, the solution must be the global optimal solution.

This property is obvious, so the proof is omitted.

2. Heuristic search algorithm

After obtaining a set of feasible solutions, we can use heuristic search methods to solve the suboptimal solution or even the optimal solution according to the characteristics of the solution. One of heuristic search methods is given as follows.

For n subtasks, assume that each subtask can match at least n candidate services in the cloud platform. Intercept the first n matching degree values $g_M(S_{ij})$ (i, j=1, ..., N,) in each ordered service set to constitute a $N \times N$ matrix

$$M = \begin{pmatrix} g_M(S_{11}) & g_M(S_{12}) \cdots & g_M(S_{1n}) \\ \vdots & \ddots & \vdots \\ g_M(S_{n1}) & g_M(S_{n2}) \cdots & g_M(S_{nn}) \end{pmatrix} \tag{9.17}$$

Based on property 1, the feasible solution can be obtained

$$\left(g_M(S_{\varepsilon_1 1}), g_M(S_{\varepsilon_2 2}), ..., g_M(S_{\varepsilon_n n}) \right) 1 \leq \varepsilon_j \leq n \; j = 1,...,n \tag{9.18}$$

The feasible solution divides the matrix m into two parts: the subscripts of the upper elements are less than ε_j, and the matching degree value is relatively large; the subscripts lower elements are greater than ε_j, and the matching value is relatively small.

$$\begin{pmatrix} g_M(S_{11}) \, g_M(S_{12})...g_M(S_{1n}) \\ ... \\ g_M(S_{\varepsilon_1 1}) \, g_M(S_{\varepsilon_2 2})...g_M(S_{\varepsilon_n n}) \\ ... \\ g_M(S_{n1}) \, g_M(S_{n2})...g_M(S_{nn}) \end{pmatrix} \tag{9.19}$$

The above matrix contains heuristic information that plays an important role in searching the better solution: (1) if the upper element can find a new feasible solution, it must be the better solution; (2) The first row of the matrix is the candidate solution of the index "optimal". In particular, when the first row is a feasible solution, it must be the global optimal solution. Based on the above heuristic information, an efficient search strategy can be designed to solve the better solution:

Heuristic strategy 1: heuristic search based on the upper element replacement. The core idea is that other services remain unchanged, and different services with large index value in a column are replaced with better services, The methods include: (1) In a certain column, the search range is from the one with high matching degree, that is, the row subscript 1 to the subscript ε_j; (2) When searching in a column, keep the feasible solution of other columns unchanged. When the first new feasible solution is found in this column or there is no new feasible solution, go to the next column; (3) when a new feasible solution is

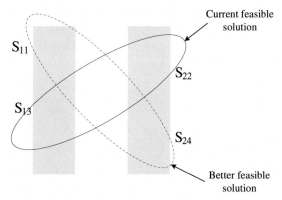

FIG. 9.2 Local optimal solution trap.

found, the larger value is retained compared with the original feasible solution; (4) The n column constitute an iteration, record the number of times to find a new optimal solution in an iteration, and terminate the iteration when there is no new optimal solution.

According to the above strategy, the better solution can be quickly found. But in the search process, only the optimal direction for the current field is considered and the local inferior solution elements are ignored, which will lead to falling into the local optimal trap. As shown in Fig. 9.2, the two services in the current solution are S_{13} and S_{22}, when there is the better solutions S_{11} and S_{24}, *i.e.* $g_M(S_{11}) + g_M(S_{24}) > g_M(S_{13}) + g_M(S_{22})$, Due to $g_M(S_{24}) < g_M(S_{22})$, S_{24} is a local inferior solution element and cannot enter the search neighborhood. In order to jump out of the trap of local optimization, the following heuristic strategy is adopted on the basis of strategy 1:

Heuristic strategy 2: heuristic search based on cross column element exchange. The core idea is that other columns of services remain unchanged, and conduct combined search for all services of a certain two columns to form a new candidate solution. The methods include: (1) Taking the columns of the two services with the worst indexes in the current solution as the search domain; (2) Judge whether the candidate solution formed by the two services in the search domain has a feasible solution. If it is a feasible solution, judge whether the index is better. If the index is better, update it to the current better solution. (3) Record the number of times to find a new better solution in an iteration, and terminate the iteration when there is no new better optimal solution.

The characteristic of strategy 1 is fast. The characteristic of strategy 2 is able to jump out of the local trap and enter the new search direction. In the comprehensive application, first quickly obtain the better solution based on Strategy 1, and then conduct deeply optimization based on Strategy 2. Record the number of better solutions obtained under the two strategies respectively, and repeat the above process. When the number of times without new better solutions after the two strategies exceeds the set value, terminate the iteration. The pseudo code flow of the above optimization search process is shown in Table 9.1.

TABLE 9.1 Pseudo codes of optimal solution searching process of parallel printing service composition.

PROCEDURE ServicesSearching ()
BEGIN ServicesSearching
      Parallelly computing the matching degrees of N subtasks. Ordering the services that can meet the requirements according to the matching degrees. Save the service data as a structure array.

      IF the N first line services different from each other THEN
            The first line is the optimal solution
      ELSE
            Compute feasible solution
            LOOP
                REPEAT
                    Compute and search the better solution according to heuristic strategy 1, compared with the solution in the previous step, record and storing
                WHILE reach the iteration termination condition of strategy 1, and record the number of new optimal solutions
                REPEAT
                    Compute and search the better solution according to heuristic strategy 2, compare it with the solution in the previous step, record and storing
                WHILE Reach the iteration termination condition of strategy 2, and record the number of new optimal solutions
                UNTIL The number of times without new better solution exceeds the limit
            END-LOOP
      END-IF
      Obtain the final optimal solution
END ServicesSearching

9.2 A 3D printing task packing algorithm

In a cloud manufacturing environment, massive 3D printing services in various types provide users with the ability of mass customization. The large number of 3D printing tasks bring more challenges on printing process management to improve 3D printers' utilization and thus saving time and materials.

This section establishes the model of 3D printing process in different types and figures out the existence of auxiliary processes. Aiming at improving the printing efficiency, an algorithm derived from the rectangle packing problem is introduced to pack printing tasks whose model size are relatively small into one task and print them all in one 3D printing process [3].

9.2.1 Problem description

3D printing technology mainly has the following categories: Fused Deposition Modeling (FDM), Selected Laser Sintering (SLS), Stereo Lithography Appearance (SLA), Digital Light Processing (DLP) and Laminated Object Manufacturing (LOM). Three of them are most commonly used: FDM, SLA and DLP.

No matter what type of 3D printer, its work is cyclic. Each work cycle includes multiple processes. All processes must go through all the processes to complete the forming of a product. There is no work process that can be merged even when working continuously.

The workflow of 3D printer can be abstracted as printing process and auxiliary processes. Printing process is the core process to form a product. The auxiliary processes provide supports to the printing process, which do not directly participate in the forming process. However, the auxiliary processes cannot be removed, otherwise the forming process cannot be executed normally, and its time consumption cannot be ignored in most cases.

Here we set the consuming time of printing process to be t_w, the consuming time of auxiliary processes be t_a, then the total time consumption of a work cycle is the sum of the two $t_w + t_a$. Define the time utilization E_c of one work cycle of 3D printer as the ratio of printing process time consumption to total time:

$$E_c = \frac{t_w}{t_w + t_a} \tag{9.20}$$

Obviously, The larger the value of E_c, the higher the proportion of printing process time in this work cycle. 3D printing working principles, 3D printer performance, working parameters and the size of task model will affect this value. The overall working efficiency *of n* work cycles $,E_p$, is the ratio of the sum of printing process time consumption and the total time in the n work cycles:

$$E_p = \frac{\sum_{i=1}^{n} t_{wi}}{\sum_{i=1}^{n}(t_{wi} + t_{ai})} \tag{9.21}$$

FDM contains four processes, heating, printing(Deposition), cooling and pick up & cleaning processes. Heating, cooling and pick up & cleaning processes are what we called "auxiliary process" for these are just pre and post treatment that do not contribute to the construction of 3D models. For the same 3D printer, the total time of auxiliary process is about a fixed value. The time of printing (deposition) process is positively correlated with the size of the 3D model to be printed. Therefore, the time utilization of one work cycle of a FDM 3D printer is positively correlated with the size of the 3D model. When printing a small size model, the sum of auxiliary processes could be larger than printing time. Continuous production of small-size models will greatly reduce the time utilization of FDM 3D printers, which should be avoided as far as possible.

SLA mainly contains three processes, printing (curing), detach, pick& cleaning. Printing is the only working process and its time consumption is determined by dedicated execution file generated by slice software. The other processes can be considered as constant time consumption. For the same 3D printer instance, the single-layer release time is fixed, and the single-layer forming time is positively correlated with the slicing area of the model. Therefore, for SLA

3D printer, the time utilization of one work cycle is positively correlated with the average cross-sectional area of the sliced 3D model and negatively correlated with the height of the model. Continuous production of "thin and tall" 3D models will significantly reduce the time utilization of 3D printers, which should be avoided as far as possible.

DLP and SLA are similar in terms of technical principles, they have same processes. The surface light source enables the DLP 3D printer to cure a whole layer of photosensitive resin at one time [4], therefore, the time consumption of monolayer curing can be regarded as a constant. The total time consumption of the printing process is only linearly related to the number of layers of the sliced 3D model. When the number of layers of a 3D model is large enough, its time utilization can be regarded as a constant.

Based on the above analysis, the overall production time can be shortened by reducing auxiliary time or improving the utilization of printing time. However, the former needs to optimize and improve the mechanical structure and electronic or electrical parts of the 3D printer itself. There are some valuable research on this area [5,6], falls outside of the scope of this book.

This section will discuss how to improve the utilization rate of printing time by effectively managing the printing process of multi printing tasks, to reduce the total printing time. Multiple 3D printing tasks from different sources will be packaged in one work cycle by using this feature of 3D printer. We proposed the concept of task package and will take FDM printer as an example to introduce 3D printing task packaging algorithm [3].

9.2.2 The 3D printing task packing algorithm (3DTPA)

1. Rectangular packing problem

In the process of 3D printing task packaging, there is an important problem to be solved, that is, the placement position of 3D model in the forming platform, which is similar to a variant of the rectangular packaging problem. The variant problem is: given a set of rectangles and a frame that can be reused and has a fixed size, put this set of rectangles into the frame and minimize the number of frames used [7].

Taking FDM printer as an example, according to the engineering practice, the task packaging problem will be studied here. There are some assumptions on the rectangular packaging problem: firstly, the scale of the problem is small, the number of tasks in a task package often does not exceed 10, and the time complexity of the algorithm is low; secondly, according to the research on the temperature field in FDM 3D printer [8], the possibility of deformation of the workpiece printed in the center of the forming platform is the lowest, so the algorithm needs to place the model to the center of the forming platform as much as possible.

The optimization of rectangular packaging problem is a NP hard problem. Heuristic algorithms are often used to solve this problem. Wu et al.

proposed a novel heuristic algorithm used in the equipment layout of drilling platform [9], which arranged the rectangle towards the center of the frame as far as possible to balance the load of the drilling platform. Referring to this heuristic solution and combined with properties of 3D printing, the 3D printing task packaging algorithm (3DTPA) is proposed [3].

2. **Definitions related to the task package**
   (1) **Task package**

   A task package is the combination of 3D printing tasks that share some attributes, such as four attributes: color, precision, layer height and material as well as auxiliary processes such as heating, cooling and release. The task package can be composed of multiple 3D printing tasks or a single 3D printing task. A 3D printer completes all printing tasks in a task package in one work cycle. Packaging the 3D printing tasks of multiple small-size models into a task package can increase the active process time in a work cycle, improve the time utilization of 3D printers, and reduce the average time-consuming of tasks.

   Multiple task packages corresponding to a 3D printer are organized using a queue model. The relationship between 3D printing task, task package and task queue is shown in Fig. 9.3. It can be seen that the tasks are not executed in the chronological order of joining the queue. We define Q_i is the task queue of the i-th 3D printer P_i, TP_{ij} is the j-th task package in the task queue Q_i of the i-th printer P_i, n_{ij} is the number of 3D printing tasks in the task package TP_{ij}.

   (2) **Layout**

   In a printing task package, a series of coordinates constitute the placement position of all printing models in a task package on the molding platform. The placement of the models is a layout. As shown in Fig. 9.4, the position of k-th model is represented by two coordinates (x_{k1}, y_{k1}), (x_{k2}, y_{k2}).

   (3) **Bounding box and enveloping rectangle**

   As shown in Fig. 9.4, the rectangle of the forming platform of the printer P_i is called the bounding box, which is denoted by Box_i. The

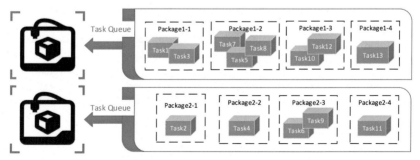

FIG. 9.3 Relationship between task, task package and task queue.

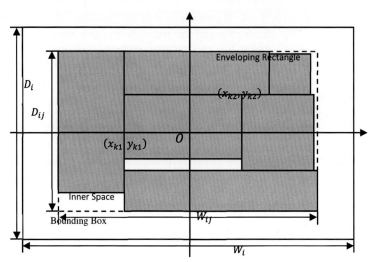

FIG. 9.4   Layout, bounding box and enveloping rectangle.

width, depth and area of the box are $W_i, D_i,$ and $Area_i$ respectively. The current task package is TP_{ij}. The smallest area rectangle (dotted line) surrounding all the current rectangular layout is called the enveloping rectangle, which is denoted by ER_{ij}. The width, depth and area of ER_{ij} are W_{ij}, D_{ij} and $Area_{ij}$ (respectively)

(4) Aspect ratio

The aspect ratio of ER_{ij} is denoted by Nar_{ij} and the aspect ratio of Box_i is denoted by Nar_i. Their values are as follows:

$$Nar_{ij} = \frac{\min\left(D_{ij}, W_{ij}\right)}{\max\left(D_{ij}, W_{ij}\right)}, Nar_i = \frac{\min\left(D_i, W_i\right)}{\max\left(D_i, W_i\right)} = \frac{\min\left(p_{i-depth}, p_{i-width}\right)}{\max\left(p_{i-depth}, p_{i-width}\right)} \quad (9.22)$$

The aspect ratio is an index to describe the narrow degree of enveloping rectangle and forming platform. The closer its value is to 0, the narrower and longer the rectangle is, and the closer it is to 1, the closer it is to the square.

(5) Filling rate

For a task package TP_{ij}, the sum of areas of all packed rectangles is denoted by $area_{ij}$. The ratio between $area_{ij}$ and $Area_{ij}$ (the area of the enveloping rectangle) is called filling rate of enveloping rectangle, which is denoted by $Efill_{ij}$. The ratio between $area_{ij}$ and $Area_i$ (the area of the forming platform) is called filling rate of forming platform denoted by $Pfill_{ij}$. They are calculated as follows:

$$area_{ij} = \sum_{M_k \in TP_{ij}} m_{k-width} \cdot m_{k-depth} \quad (9.23)$$

$$Efill_{ij} = \frac{area_{ij}}{Area_{ij}} = \frac{area_{ij}}{D_{ij} \cdot W_{ij}}, \; Pfill_{ij} = \frac{area_{ij}}{Area_i} = \frac{area_{ij}}{p_{i-width} \cdot p_{i-depth}} \qquad (9.24)$$

High filling rate means that the space of enveloping rectangle and forming platform is more fully utilized. However, in order to avoid the influence of uneven temperature and edge deformation at the edge of the forming platform on the forming process of 3D model, the edge part must be cleared and the filling rate cannot be blindly improved. Generally, 65% is an appropriate value of the filling rate.

(6) Inner space

As shown in Fig. 9.5, in the packaging process, many blank spaces will inevitably appear inside the packaging rectangle. Place a virtual rectangle in the blank space and expand the area of the rectangle until the edge of the virtual rectangle coincides with the edge of the packed rectangle or the edge of the enveloping rectangle. The space occupied by this virtual rectangle is called inner space. There are two inner spaces in Fig. 9.5, there are overlapping parts between them. In the 3D printing task packaging algorithm, the analysis of inner space is a very important part.

(7) The average height of models in one task package

Generally, there are several task models in one task package. The average height of them is denoted by $\bar{h}_{ij}$ and it is calculated as follow:

$$\bar{h}_{ij} = \frac{\displaystyle\sum_{M_k \in TP_{ij}} m_{k-height}}{n_{ij}} \qquad (9.25)$$

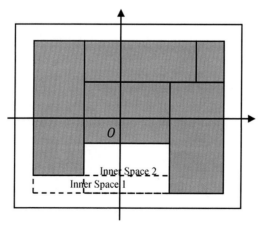

FIG. 9.5 Inner space.

3. Matching degrees of 3D printing models and 3D printers

Another major part of the 3DPA algorithm is to calculate the matching degrees between 3D models and printers. The definitions of matching degrees of several typical parameters have been given in Chapter 8. They are the type of printers, printing accuracy, printable space, supportable document types, types of materials, and colors of materials. However, for the task packaging algorithm, different aspects need to be considered, so the parameters that need to be matched are also different. Six parameters are selected as shown in Table 9.2.

Where, matching degrees of printing accuracy, types of materials, and colors of materials are defined in Section 8.2.1, the matching degrees of other parameters are defined as follows.

(1) Matching degree of bottom size. When placing the 3D model in the print task on the forming platform, first consider whether the inner space is enough to place the model to place the model; If not, the model will be placed to the edge of enveloping rectangle. If there is still not enough space, create a new task package and place the 3D model in the center of the forming platform.

The calculation method of the matching degree changes with the position where the 3D model is placed. The best solution is to place the 3D model in the internal space, which will not increase the area of the enveloping rectangle, and can increase the filling rate of the enveloping rectangle **Efill**$_{ij}$ and the filling rate of the forming platform **Pfill**$_{ij}$, so as to make better use of space. For this situation, the matching degree is

$$g_i\left(m_{k-width}, m_{k-depth}\right) = 1 \qquad (9.26)$$

TABLE 9.2 Six parameters of 3D printers and tasks.

Dimension	3D printer	3D printing task
Size (width, depth)	Size of molding platform	Size of 3D model
Height	Height of molding space	Height of 3D model
Color	Material color, an element of the color set	Acceptable material colors, an subset of the color set
Accuracy	Extruder diameter (FDM), laser diameter (SLA), projector pixel size(DLP)	Minimum acceptable accuracy of the 3D printer
Layer height	Minimum layer height that a 3D printer can print	maximum layer height of the production.
Material	Material type selected from {ABS, PLA, TPU, UVCR, FUVCR}	Material requirement from {flexible, inflexible}

When the 3D model is placed adjacent to the edge of the enveloping rectangle, the area of the enveloping rectangle will increase, which is not the best solution. Since the 3D model can rotate 90° around its z axis during placement, the placement schemes can be divided into four categories. At this time, we hope that the area increment of the enveloping rectangle is as small as possible. The matching degree is as follows:

$$g_i\left(m_{k-width}, m_{k-depth}\right) = \max\left\{\frac{m_{k-width}}{W_{ij}}, \frac{m_{k-width}}{D_{ij}}, \frac{m_{k-depth}}{W_{ij}}, \frac{m_{k-depth}}{D_{ij}}\right\} \quad (9.27)$$

When the 3D model is placed in the center of the forming platform, we want the aspect ratio to be as close as possible to the forming platform, and the filling degree to be as high as possible. The matching degree is as follows:

$$g_i\left(m_{k-width}, m_{k-depth}\right) = \frac{\min\{Nar_k, Nar_k\} \cdot area_k}{\max\{Nar_k, Nar_i\} \cdot Area_i} \quad (9.28)$$

(2) Matching degree of height. Assuming that there are two 3D printing tasks in the same task package, and there is a huge gap in the height of their 3D models, the following situations may occur: the lower height model has been finished, but the higher height model is still being printed, now the printed model cannot be removed from the forming platform, they can only be removed after all the models are printed and the work cycle enters the part taking and cleaning phase. Under the extreme conditions that there is one extremely high model and multiple low models in a task package, the completion time of the low model will be greatly delayed. Therefore, we tend to put print tasks with similar model height into the same task package. The matching degree of height is as follows:

$$g_i\left(m_{k-height}\right) = 1 - \frac{\left|m_{k-height} - \bar{h}_{ij}\right|}{\max\left\{m_{k-height}, \bar{h}_{ij}\right\}} \quad (9.29)$$

When $\bar{h}_{ij} = 0$, that is to put a 3D model into a new task package,- $g_i\left(m_{k-height}\right) = 1$.

(3) Matching degree of layer height. For a 3D printer, the layer height refers to the minimum height of each layer of the sliced model that can be printed by the 3D printer. The matching degree is as follows:

$$g_{ij-layerheight} = \begin{cases} 1 & m_{k-layerheight} \geq p_{i-layerheight} \\ 0 & m_{k-layerheight} < p_{i-layerheight} \end{cases} \quad (9.30)$$

(4) Total matching degree. The total matching degree is the product of the matching degrees of all parameters, which is calculated as follows:

$$G_{ij} = g_{ij-size} \cdot g_{ij-height} \cdot g_{i-color} \cdot g_{i-accuracy} \cdot g_{i-layerheight} \cdot g_{i-material} \quad (9.31)$$

4. Steps of the algorithm

Step 1: Calculate matching degree of each task for each 3D printer without packing them, the number of matching degrees for each task is the same as the number of 3D printers. Sort the tasks in descending order of bottom area of each model.

Step 2: For a task, traverse all existing task packages and calculate its matching degree for all available task packages, then find the appropriate place in each available task package for the task.

Step 3: Select the task package that get the highest fit degree and put the model in the appropriate place on the molding platform.

The key of the algorithm is to find the appropriate placement position of the 3D model in the forming platform. As shown in Fig. 9.6, When model M can be placed in the inner space, it should be placed in the corner of the interior space, and the overlapping edge of model M and the existing model should be as long as possible. Scheme 1 is the best, scheme 2 is the second, scheme 3 is the third and scheme 4 is the fourth.

As shown in Fig. 9.7, when the model M needs to be placed adjacent to the enveloping rectangle, make the overlapping edge of the model M with

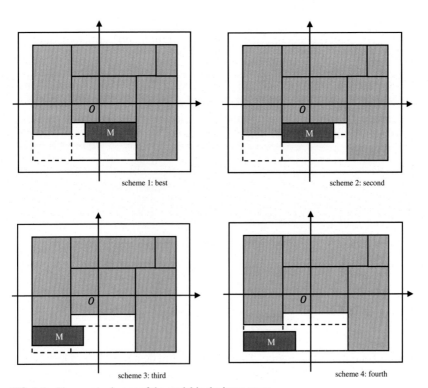

FIG. 9.6   Placement schemes of the model in the inner space.

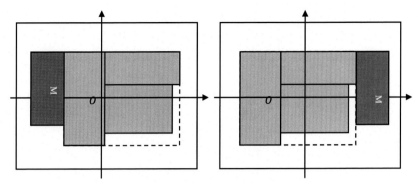

FIG. 9.7 Placement scheme of the model in the adjacent enveloping rectangle.

the existing model and the new enveloping rectangle as long as possible, and this position is the best one.

5. Case study

Based on the working principles of 3D printers and slicing software tools developed by major 3D printer manufacturers, such as OctoPrint, Cura, PreForm and NanoDLP, we can estimate the time consumed in one work cycle of various 3D printers. According to the parameters of 3D printers obtained from the official websites of major 3D printer manufacturers, and records of working history of actual 3D printers, various virtual 3D printers can be generated. In this experiment, five virtual 3D printers are generated. Parameters of these virtual printers are listed in Table 9.3.

3D models are selected from the model library of the cloud 3D printing platform, and task parameters are randomly generated to generate 3D printing tasks, with a quantity of 30 pieces. The 30 3D printing tasks are packed according to the algorithm and assign them to 5 virtual 3D printers. The results are shown in Table 9.4. The layout of a typical task package TP_{33} is shown in Fig. 9.8.

Use relevant software to estimate the time spent by each 3D printer to complete the printing task. Five virtual 3D printers are set to work in parallel, and the maximum time is the total time to complete all tasks. Through the simulation experiment, the time spent by each 3D printer to complete the task under the condition of packaging and unpackaging is compared, and the results are shown in Fig. 9.9.

According to the experimental results, it can be calculated that the total time is reduced by about 8.5%. For each 3D printer instance, the proportion of time savings is different, with the maximum reaching 60%. Through in-depth analysis of the packaging results, it can be found that the difference in time saving proportion is mainly related to the type of 3D printer and the size of the 3D model in the task package. Task packaging is most effective for DLP 3D printers. When packaging multiple 3D models into the same task

TABLE 9.3 Parameters of virtual 3D printers (mm).

Name	P_{width}	P_{depth}	P_{height}	P_{color}	$P_{accuracy}$	$P_{layerheight}$	$P_{material}$
Ultimaker (FDM)	220	210	230	Red	0.4	0.1	PLA
PrintRite (FDM)	280	180	180	Yellow	0.3	0.2	TPU
RepRap (FDM)	180	100	100	Black	0.5	0.2	ABS
Xiaofang (SLA)	130	130	180	Blue	0.1	0.05	FUVCR
ProjectD (DLP)	100	76	150	White	0.05	0.05	UVCR

TABLE 9.4 Packaging results of 3D print tasks.

Names of 3D printers	Task package 1	Task package 2	Task package 3
ProjectD	14, 2, 4	Null	Null
Cubic3D	21, 17, 19	Null	Null
RepRap	1, 16, 7, 15	20, 3, 22, 9, 13, 12	30, 6, 26, 23, 25, 8
PrintRite	28, 29, 24	27	Null
Ultimaker	5	18, 11, 10	Null

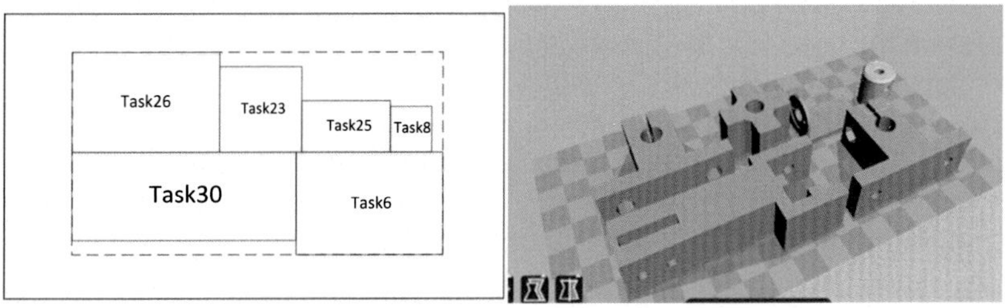

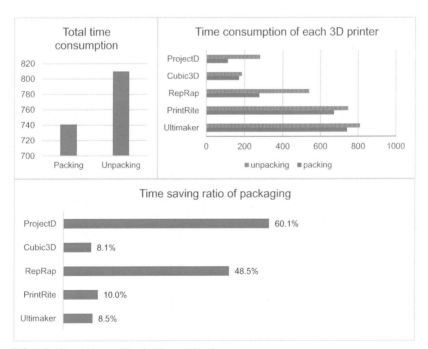

FIG. 9.8 The package of a 3D printing task.

FIG. 9.9 Simulation results of different 3D printers.

package, for FDM and SLA 3D printers, the printing time is the sum of the time of each model being printed separately. But for DLP printers, the printing time is equal to the longest time of each model being printed separately, which result in more obvious effect of time saving. In addition, when the size of 3D models in the task package is small and the number is large, the effect of task packaging on saving time is also very obvious, because when the small-size 3D model is printed alone, the time utilization of the work cycle is low, so that the effect of task packaging on improving the time utilization of the work cycle is more significant.

9.3 Conclusion

This chapter focuses on how to improve the printing efficiency through the management of the printing process when the printing platform processes multiple printing tasks. For the printing task of complex products, it is necessary to decompose the printing task of complete products into printing tasks of multiple parts and complete them in parallel. A parallel processing service composition optimization method based on heuristic search algorithm is proposed, which can make full use of the rich 3D printing services in the cloud platform and significantly improve the efficiency of multitask printing. There are a lot of research on service composition methods in cloud manufacturing related research [10–15]. As a special manufacturing service, 3D printing service composition can make full use of these methods.

Another way to improve printing efficiency is to make full use of the working principle of 3D printers, reasonably manage multiple printing models, and increase the proportion of printing time in the whole production time to improve the overall printing efficiency of multiple models. The proposed 3D printing task packing algorithm in this chapter packs tasks into one package and build them in the same printing cycle. The approach of packing models is based on rectangle packing solutions and 3D printing constraints. More dimensions and details of 3D printer and 3D printing service can be considered with respect to different printing requirements in applications of the algorithm.

References

[1] K.C. Lu, H.M. Lu, Combinatorial Mathematics (in Chinese), Tsinghua University Press, 1991.

[2] R.A. Brualdi, Introductory Combinatorics, fourth ed., Prentice Hall, 2004.

[3] Z. Zhao, L. Zhang, J. Cui, A 3D printing task packing algorithm based on rectangle packing in cloud manufacturing, in: *Proceedings of Chinese Intelligent Systems Conference*, Mu Dan Jiang, China, Springer, 2017, pp. 21–31.

[4] G.H. Wu, S. Hsu, Review: polymeric-based 3D printing for tissue engineering, J. Med. Biol. Eng. 35 (3) (2015) 285–292.

[5] J.R. Tumbleston, D. Shirvanyants, N. Ermoshkin, et al., Continuous liquid interface production of 3D objects, Science 347 (6228) (2015) 1349–1352.

[6] K.Y. Fok, N. Ganganath, C.T. Cheng, et al., A 3D printing path optimizer based on Christo-fides algorithm, in: Proceedings of IEEE International Conference on Consumer Electronics-Taiwan (ICCE-TW), IEEE Xplore, 2016, pp. 1–2.

[7] R.E. Korf, M.D. Moffitt, M.E. Pollack, Optimal rectangle packing, Ann. Oper. Res. 179 (1) (2010) 261–295.

[8] J.L. Cao, Numerical Simulation of Temperature Field and Stress Field of FDM Rapid Proto-typing Machine, Harbin Institute of Technology, 2014.

[9] L. Wu, L. Zhang, W.S. Xiao, et al., A novel heuristic algorithm for two-dimensional rectangle packing area minimization problem with central rectangle, Comput. Ind. Eng. 102 (2016) 208–218.

[10] H. Guo, L. Zhang, Y.L. Liu, et al., A discovery method of service-correlation for service com-position in virtual enterprise, Eur. J. Ind. Eng. 8 (5) (2014).

[11] Y.K. Liu, X. Xu, L. Zhang, F. Tao, An extensible model for multi-task oriented service com-position and scheduling in cloud manufacturing, J. Comput. Inf. Sci. Eng. 16 (4) (2016), 041009.

[12] F. Li, L. Zhang, Y.K. Liu, Y.J. Laili, F. Tao, A clustering network-based approach to service composition in cloud manufacturing, Int. J. Comput. Integr. Manuf. 30 (12) (2017) 1331–1342.

[13] F. Li, L. Zhang, Y.K. Liu, Y.J. Laili, QoS-aware service composition in cloud manufacturing: a Gale-Shapley algorithm-based approach, IEEE Trans. Syst. Man Cybern. Syst. 50 (7) (2020) 2386–2397.

[14] F. Wang, Y.J. Laili, L. Zhang, A many objective memetic algorithm for correlation aware ser-vice composition in cloud manufacturing, Int. J. Prod. Res. 59 (17) (2021) 5179–5197.

[15] H.G. Liang, X.Q. Wen, Y.K. Liu, H.F. Zhang, L. Zhang, L.H. Wang, Logistics-involved QoS-aware service composition in cloud manufacturing with deep reinforcement learning, Robot. Comput. Integr. Manuf. 67 (2021) 101991.

Chapter 10

Security and privacy in cloud 3D printing

10.1 Data security of cloud 3D printing platforms

With the continuous development of cloud manufacturing platform technology, security issues have become an important bottleneck hindering the development of cloud manufacturing. Since the cloud manufacturing platform needs to support two-way communication between the manufacturing service demander (platform customer) and the manufacturing service provider (manufacturing enterprise), the security risk of the cloud manufacturing platform is increased. With the increase in transaction volume in cloud manufacturing platforms, the security threats to user data and manufacturing equipment in cloud platforms also increase significantly. The security problems of cloud 3D printing platforms are mainly reflected in three aspects: data security, 3D printing equipment access control and credit security problems.

10.1.1 Data security issues in cloud 3D printing

Data security is one of the key issues hindering the development and application of cloud manufacturing. In order for cloud manufacturing to be truly applied and implemented on a large scale in the manufacturing industry, the security issues in cloud manufacturing must be resolved. A reliable cloud manufacturing platform should adopt effective cloud security mechanisms to ensure the security of the cloud platform, including cloud data security, manufacturing equipment access control, network security, etc. Some common security technologies that can be used to enhance the security of cloud manufacturing platforms and reduce security risks are: data encryption technology, virtual local area network technology (which can provide secure remote communication), network middleware (such as firewalls, packet filters, etc.). There are currently a few companies that have designed and developed software and tools specifically for cloud platform security, but these software are often only suitable for cloud computing environments. And data security and privacy management in cloud manufacturing is still in its early stages. The service provider in the

Customized Production Through 3D Printing in Cloud Manufacturing
https://doi.org/10.1016/B978-0-12-823501-0.00013-4

cloud manufacturing platform, that is, the manufacturer pays more attention to the protection of its data in the cloud manufacturing platform. When different business entities or manufacturers use cloud manufacturing platforms to share data, these data usually need to be transmitted through the Internet and some other information systems. At this time, how to protect the security of enterprise data is a very challenging task for cloud manufacturing platforms. Because the cloud manufacturing platform involves a lot of sensitive data of the enterprise, such as data about customers, employees, business processes, etc., the cloud manufacturing platform must deploy strong security mechanisms to enhance the security of the data in the cloud manufacturing platform [1].

As noted in a recent Cloud Security Alliance (CSA) report, data breaches are one of the most common security hazards [2]. And from a legal and economic point of view, data breaches have proven to have enormous adverse consequences. In terms of retrieving data and modifying data, cloud data breaches refer to unauthorized access to data hosted in cloud platforms. This is a particularly important issue in cloud manufacturing platforms. A data breach of a cloud manufacturing platform can result in the loss of sensitive business information of client companies, such as trade secrets or contract details. This kind of corporate information leakage problem can have a serious negative impact on the company. Encryption is the most common solution to preventing a data breach, but sometimes encrypting data in a cloud manufacturing platform doesn't fully solve the data breach. Because in the context of cloud manufacturing, the origins of data breaches are usually internal to the company rather than external. As the CSA report points out, the fact that company insiders (such as current or past employers, system administrators) and company partners (such as contractors and business partners) may be more likely to be the culprits of company data breaches [2]. A common data protection measure for cloud manufacturing platforms is the use of customer-controlled encryption keys. But this way is ineffective against malicious insiders who might have the correct decryption key. Therefore, manufacturing service providers (i.e., manufacturing enterprises) that use cloud manufacturing services also need to adopt advanced key management solutions to achieve further data security assurance.

10.1.2   Encryption technology for cloud 3D printing data

In cloud manufacturing systems, vulnerabilities in key management are one of the main factors leading to data breaches, as company insiders can obtain valuable documents and trade secrets related to company products and manufacturing processes and maliciously share them with competitors, or use these materials to start your own business. Preventing such breaches requires a key management system in place to record which employees hold certain keys and to revoke those keys when employees don't need them. The key is to have strict key management to reduce the possibility of breaches. But it must be premised that employees can effectively complete their work, and it should not be

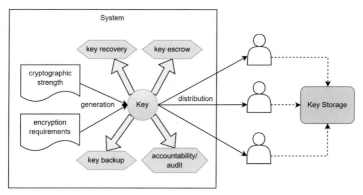

FIG. 10.1 Phases of key management.

overkill. Despite all the precautions companies take, data breaches can still happen. Manufacturing companies in cloud manufacturing platforms must be able to detect data breaches in a timely manner to prevent malicious insiders and competitors from further exploiting the vulnerability to gain access to company documents and secrets.

A possible and effective way to prevent data leakage in cloud manufacturing is to develop a reasonable key management scheme. And in the key management scheme, effective technical schemes are formulated for each stage in the process. Fig. 10.1 shows a possible implementation of key management for different stages [3].

(1) Generate keys using cryptographic modules that meet cryptographic strength and encryption requirements.
(2) Distribute keys to all authorized users using a valid key distribution method.
(3) Users can choose to store keys in a secure key store.
(4) For systems with long-term data at rest, a key recovery plan needs to be configured to prevent key loss.
(5) Record each user who receives or controls the key to ensure traceability when the key is compromised.
(6) When a key is compromised, revoke the key and generate a new key, which is then sent to authorized users.

10.1.3 Violation identification and notification in cloud 3D printing

A key management strategy is only the first line of defense against cloud-manufactured data breaches. Due to the complexity of cloud manufacturing systems involving a large number of enterprises and users, data leakage is difficult to avoid. Therefore, cloud manufacturing systems must be equipped with an effective method to identify and record breaches related to data breaches,

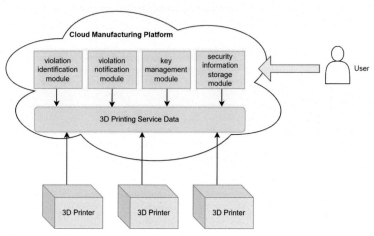

FIG. 10.2   A possible solution for data protection in cloud manufacturing.

and notify relevant personnel of these breaches in a timely manner. Specifically, cloud manufacturing requires the ability to prevent and identify data security-related breaches (such as data being read or altered by unauthorized individuals) and to notify data owners in cloud manufacturing platforms of such breaches. In addition, the cloud manufacturing platform should also have the ability to collect and store information related to data security breaches in a lawful manner. An effective cloud manufacturing data breach solution should support the four stages of data breach prevention, identification, notification, and recording. Such solutions can be deployed as software-as-a-service (SaaS) for easier integration of data protection software tools into the operational processes of cloud manufacturing platforms. Fig. 10.2 shows a feasible SaaS solution for preventing data leakage in cloud manufacturing, including a data security violation identification module, a key management module, a data security information storage module, and a data security violation notification module. Violation identification related to data security in cloud manufacturing systems is still a cutting-edge research topic. There is still a lack of very effective cloud manufacturing data security solutions. One possible solution is to use digital watermarking or other techniques to hide encryption on data in cloud manufacturing systems. These techniques and methods need to be able to detect illegal data security breaches such as data breaches and malicious data access.

10.2   Access control for cloud 3D printers

The security problem of cloud manufacturing platform is not only reflected in the protection of data, but also in the protection of manufacturing equipment, that is, to avoid malicious operations such as illegal access and control of the manufacturing equipment of service providers through the Internet and cloud platforms. In order to ensure the security of remote access to 3D printing equipment in the cloud manufacturing platform, it is necessary to ensure that only personnel with corresponding access rights can remotely access and operate 3D printers. Cloud manufacturing platforms need to focus on and control the permissions of different users to remotely access and control 3D printers [4]. One possible way to control remote access to 3D printing devices is to enhance the security of cloud 3D printers by providing exclusive access to the resources of cloud virtual machines and restricting access by unauthorized users [1]. To enable authorized users to legally access cloud 3D printing resources at the right time, geographic location, access time, and subnet information can be mapped to the key's geo-authentication model. In addition, a conflict firewall can be set in the cloud manufacturing platform to isolate cloud manufacturing users with access conflicts.

Since the architecture and resource management mechanism of cloud manufacturing are different from those of cloud computing, the security mechanisms in cloud computing cannot be directly used in cloud manufacturing platforms. Cloud security has become a concern for many manufacturing companies. Cloud-based services have a higher risk of unauthorized access (unauthorized access) and malicious access. The reason why device access control in cloud manufacturing platforms faces great challenges is that cloud manufacturing platforms need to support a large number of users accessing centrally managed manufacturing resources at the same time. Cloud manufacturing adopts a multi-tenant approach to realize the sharing of manufacturing resources by different users.

These tenants in the cloud manufacturing platform have the ability to design and develop their own software, and share the software they developed as a service to other cloud manufacturing users through the cloud manufacturing platform. The software services developed by tenants may have certain defects, and these defects are likely to cause malicious attackers to illegally access and invade manufacturing resources and equipment through the cloud manufacturing platform. And different tenants in a cloud manufacturing platform may often use different virtual machines on the same physical server. In this case, the security mechanism used to restrict access to manufacturing resources and control user permissions is more important, because malicious attackers have the opportunity to use this mechanism to illegally access data, resources and other users shared on the same physical machine. Material. These malicious behaviors are important security risks for cloud 3D printing platforms. During the operation of the cloud 3D printing platform, 3D printing manufacturing resources are frequently accessed and invoked. In this process, the security of data files and models related to cloud 3D printers is also very important, such as important processing task data, product design files, files of 3D printing models, etc.

In addition to the resource access restriction and user access control of the cloud 3D printing platform, the method of data confidentiality protection can also be used to solve the security problem of the 3D printing equipment in the cloud 3D printing platform. Here we give an example to illustrate the necessity of this data protection technology. Suppose a 3D printing task order is received in the cloud 3D printing platform. There are two manufacturing companies A and B in the cloud 3D printing platform bidding for this order. Due to the needs of the processing technology, both companies A and B need to send the processed 3D printed product prototypes to manufacturing company C for the next step of product surface treatment, because company C has unique and expensive equipment to carry out special 3D printed product surfaces treatment process. Assume that enterprise A has completed the development of the product prototype, and after starting the virtual machine of enterprise A in the cloud platform, the related files of the product prototype designed and developed by it are sent to enterprise C through the cloud manufacturing platform. Since the surface treatment process of the product takes a certain amount of time, A chooses to shut down its virtual machine after C successfully receives the product prototype file to save resource consumption costs. At this point, Company B is also bidding for the same contract and is also prototyping the product. Therefore, enterprise B also chooses to send the related documents of the product prototype designed and developed to C through the cloud manufacturing platform. At this time, enterprise B needs to communicate with enterprise C and has permission to wait list of its jobs. However, at this time, the job of enterprise A is also on the waiting list. Without strict resource access controls, Enterprise B could potentially snoop on Enterprise A's prototyping-related files. These documents of Enterprise A are very sensitive confidential documents for bidding and tendering. This behavior of enterprise B endangers the confidentiality of enterprise A's data. In order to avoid this security risk, the methods of resource access control and user permission control can be used to improve the security of user resources in the cloud 3D printing platform. In order to improve the security of the device in the cloud 3D printing platform, the user's geographic location, the user's access time, and the user's network area can be considered in the encryption mechanism. In the design of the key, the decryption area, decryption time and the information of the decryption network are considered accordingly to improve the security of the key. Specifically, the information recipient must be located in a specific decryption area in space, within a specific decryption time in time, and in a specific decryption network in order to decrypt the message and access 3D printing resources in cloud manufacturing. In addition, the conflict firewall technology in the manufacturing cloud can also be used to identify cloud manufacturing users with conflicting interests. At the same time, users on different virtual machines in the same physical machine can be distinguished. The resources of cloud manufacturing users are valuable assets that cloud manufacturing platforms need to protect. The main security threats that need to be considered in the resource protection mechanism include information leakage, information tampering, activity denial, and masquerading.

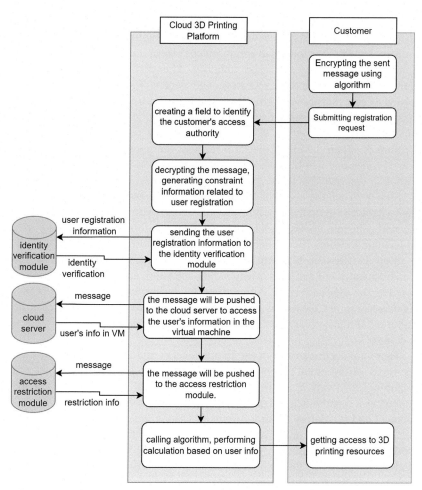

FIG. 10.3 System framework of cloud 3D printing resource access control.

Fig. 10.3 shows the System framework of cloud 3D printing resource access control. Cloud manufacturing platform users need to register their physical location for subsequent resource matching and scheduling. In the cloud 3D printing resource access control system, there is a Cloud Gate Access Control module which includes a user authentication module and an authentication module.

Step 1: The user encrypts the sent message using an algorithm. Before submitting the documents, the user submits a request to register an account. Different types of user information are signed by the user's private key. After that, the message is further encrypted and protected. After the account registration is completed, users will have permission to publish their own resources or access public resources on the cloud platform.

Step 2: After the cloud platform receives the user's message, it will create a field to identify the user's access authority. After receiving the message, the cloud platform decrypts the message and generates constraint information related to user registration. After the cloud platform verifies the user information, it will send the user registration related information to the identity verification module.

Step 3: When the cloud platform receives the request to authenticate the user, the message will be pushed to the cloud server to access the user's information in the virtual machine in the cloud platform.

Step 4: When the cloud platform receives the response message, the message will be pushed to the access restriction module. The access restriction module will schedule corresponding user restriction information, including user geographic location, access time and user subnet information.

Step 5: After the platform user receives the encrypted information, the cloud platform will call the corresponding algorithm and perform calculation based on the user's location, current time and its subnet address. If the access conditions are met, the user is granted access to the resource and can access the resource. If the access conditions are not met, the user's access request is denied.

There are security risks in the remote access and control of 3D printing resources in the cloud 3D printing platform. At present, the methods and technologies in this area are still relatively limited. This problem can be defined as a communication security problem, or it can be defined as a management and regulatory security problem. During the operation of the cloud manufacturing platform, corresponding security measures must be strengthened to minimize risks. A key issue in implementing a cloud manufacturing platform is ensuring that platform users' proprietary data and intellectual property are adequately protected. Moreover, the cloud manufacturing system needs to meet the data privacy protection of platform users and enterprises. The purpose of privacy protection is to improve data security and control the access rights of different users to data and resources. During the operation of the cloud manufacturing platform, the protection of user data and resources can improve the security of physical manufacturing resources and operators related to manufacturing equipment in the cloud manufacturing environment. But most of the current research pays more attention to data security and device security. Research and discussion focusing on the safety of equipment operators is relatively limited. Security measures in the cloud manufacturing platform are a necessary condition to protect the security of data, software resources, hardware manufacturing resources, and user privacy in the cloud platform [4].

10.3   Security of cloud 3D printing based on blockchain

10.3.1   Review of blockchain applications in cloud manufacturing

With the development of cloud manufacturing technologies, the lack of trust among different users on the platform has become one of the biggest bottlenecks. In recent years, blockchain (BC) technology was proposed and applied in different industries. The concept of BC was firstly proposed by Satoshi Nakamoto in 2008 [5], which can be understood as an immutable distributed ledger protected by a specific encryption method. Advantages of BC include trustlessness, tamper resistance, traceability, and high transparency. According to scopes, BC can be divided into three categories: public chains, alliance chains, and private chains. Table 10.1 shows the differences between these categories.

In addition to the financial industry, BC has also been applied in medicine, agriculture, food, and manufacturing. The possibility of applying BC in cloud manufacturing also attracted attention. BC could establish a way of interaction and trust between different entities in cloud manufacturing platforms through a distributed network structure.

BC-based distributed cloud architecture can provide a secure and trusted data-sharing solution for cloud manufacturing platforms [6]. The injection mold redesign and manufacturing knowledge sharing system based on cloud manufacturing and BC technology can use BC technology to record transaction data to ensure the security and reliability of cloud manufacturing systems [7]. The development and implementation of cyber-physical production systems using BC technology in the real world will have a positive impact on the operation of cloud manufacturing platforms [8]. Decentralized cloud manufacturing applications were developed through BC-based smart contract technology to achieve more transparent, economical, and secure cloud manufacturing protocols [9]. On one hand, BC can be applied to the workshop and manufacturing equipment level of cloud manufacturing systems [10]. Adding BC technology at the workshop level can realize data sharing within the workshop. BC can play a special role in quality assurance [11]. The data generated in the smart manufacturing process can be used to retrieve material sources, improving equipment management efficiency and transaction efficiency. On the other hand, BC can also be applied to platform level for cloud manufacturing. By adding BC-based service composition technologies, users can directly see the default rate of service providers, and thus choose whether to conduct transactions without relying on third-party platforms [12]. The matching and transaction process of manufacturing services based on smart contracts is also another important application [13]. In addition, BC-based data recording technologies can realize the recording, storage, and sharing of transaction data in the cloud manufacturing platform, and ensure the security and reliability of the cloud manufacturing service platform information [14]. The BC-based cyber-physical system was used as a reliable network to eliminate the trust problem of the third

TABLE 10.1 Three categories of block chains.

Categories	Writing/reading permission	Pros	Cons.	Applications
Public chains	Public/public	High decentralization and transparency	Low privacy	Bitcoin, Ethereum
Alliance chains	Semi-public/semi-public	Low maintenance cost, high node reliability	Low transparency	Hyperledger Fabric, R3 Corda, BCOS, Hyperchain
Private chains	Private/semi-public	High privacy	Low decentralization and transparency	

party in the cloud manufacturing platform, thereby improving the security of small enterprises [15].

At present, applications of BC in cloud manufacturing are mainly concentrated at the system architecture level [6]. Detailed customized development of BC in specific cloud manufacturing platforms needs further discussions. Particularly, more studies about the consensus mechanisms of BC in cloud manufacturing need to be done. The system architecture of BC is shown in Fig. 10.4. The consensus algorithms are the core and especially important for applying BC in cloud manufacturing.

Due to the complexity of cloud manufacturing systems, it is difficult for classical BC consensus algorithms to fully meet the actual needs in cloud manufacturing systems. Therefore, it is valuable to design specific BC consensus algorithms for specific cloud manufacturing scenarios. At present, most cloud manufacturing platforms are still using typical consensus mechanisms such as PoW and PBFT [16,17] which lack customization for cloud

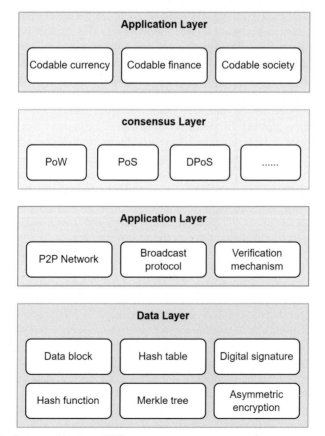

FIG. 10.4   System architecture of BC.

manufacturing scenarios. The consensus mechanism is the core of a BC network. Designing a good consensus algorithm would greatly affect BC's transmission performance and stability.

10.3.2 Credit security of cloud 3D printing services

Focusing on the real cloud manufacturing scenarios and demands, aiming at the problems of token dependence, low efficiency and high resource consumption in the current blockchain consensus mechanism, a consensus mechanism of service capability proof for cloud manufacturing is proposed, including the service capability calculation method, the trust incentive mechanism, and the service capability proof algorithm.

1. **Service capability calculation method**

Service capability (SC) is an indicator that describes the SC of member nodes in the cloud manufacturing service supply chain network at the current moment, which can quantify and characterize the SC of member nodes in the cloud manufacturing service supply chain. The member node is measured by the historical transaction volume, total transaction value and user evaluation obtained in the cloud manufacturing service supply chain system. Specifically, the SC is the product of the historical transaction volume of each member node, the time factor and the user evaluation factor.

The core of service capability proof lies in the calculation of the SC of each service provider in the cloud manufacturing system. For the SC of cloud service providers, not only the current latest transaction data, but also its historical transaction data need to be considered. At the same time, different weights need to be assigned to transaction data in different periods. On the other hand, user evaluations also need to be considered. To reflect the service quality and delivery punctuality rate of the service provider, so as to measure and evaluate the SC of the service provider from a global perspective. SC can be calculated through the following formula.

$$SC = \sum_{m=(I_{last}-N+1)}^{I_{last}} \left(Tx_m \times \frac{1}{1-(I_m - I_{last})} \right) \qquad (10.1)$$

SC represents the SC of service providers. Tx_m represents the total amount of cloud manufacturing service transactions recorded by this service provider in block m. $\frac{1}{1-(I_m - I_{last})}$ represents the time factor of the total amount of service transactions of the service provider in the m block, which reflects the different weights of the SC formula on transaction data in different periods. And m represents the m-th block in the service chain; I_{last} represents the height of the end block in the service chain; I_m represents the height of the m-th block in the service chain; $I_m - I_{last}$ represents the mth block in the service chain The height difference between the block and the end block, because the height of the block is related to time, it can represent the time sequence when the block is generated to a certain extent, so this parameter can represent the time difference between the blocks to a certain extent, and The block height is an integer, which can make the quantification of the time difference more standard; N represents the number of blocks in the service chain that needs to be inspected to calculate the SC of the cloud service provider; $I_{last} - N + 1$ represents the selected inspection The initial block of blocks in the N service chains of. Among them, the calculation formula of the total service transaction Tx_m of a service provider in block m is as follows:

$$Tx_m = \sum_{i=1}^{n}(r_i \times Tx_i) \tag{10.2}$$

In the formula: Tx_i represents the total price of the ith transaction recorded by the service provider in block m; r_i represents the evaluation factor of the ith transaction; n represents the price of the service provider in block m. The total number of transactions recorded. Among them, the calculation formula of the evaluation factor r_i of the transaction is as follows:

$$r_i = \frac{Score_i}{Score_{max}} \tag{10.3}$$

$Score_i$ represents the user evaluation score of the transaction. $Score_{max}$ represents the user's full score when evaluating the transaction. For the time factor in Eq. (10.1), since I_{last} represents the height of the end block in the service chain, and I_m represents the height of the m block in the service chain, there is always $I_{last} \geq I_m$, which means $(I_m - I_{last}) \leq 0$. Let the parameter $t = (I_m - I_{last})$, then the function expression of the time factor can be obtained as:

$$F_t = \frac{1}{1 - t} \tag{10.4}$$

t is a negative integer and $t \in (-\infty, 0]$. The function of the time factor is shown in Fig. 10.5.

Through the transformation of the above time factor function, the height information of the block is mapped into the interval $(0,1]$. And the closer it is to the end block, the closer the time factor is to 1. On the other hand, no matter how far a block is from the end block, it can be guaranteed that a positive value will be obtained after the transformation of the time factor function. Therefore, the corresponding transaction can contribute to the SC of the service provider. Through the transformation of the above time factor function, The SC formula can fully update the development status of the enterprise, not only can use the historical transaction data to judge the enterprise's previous service ability, but also can use the latest transaction data to judge the improvement of the enterprise's recent service ability.

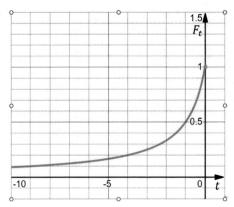

FIG. 10.5   Function of the time factor.

For the evaluation factor in Eq. (10.1), the calculation of *SC* not only considers the objective factors such as the order volume and transaction amount that the service provider can obtain, but also introduces the more subjective factor of user evaluation. It can better reflect the enterprise's service ability in multi-service parallel processing ability, service delivery punctuality rate and publicity and promotion, and further improve the service ability formula to evaluate the service ability of service providers.

The service transaction information used for service force calculation is stored in the service chain. The service chain records the specific information of each service transaction in the cloud manufacturing supply chain network. Since the transaction information of the cloud manufacturing service is stored in the service chain, the service transaction data is endowed with the characteristics of non-tampering and traceability. After that, it is necessary to broadcast the whole network and upload the transaction data to the chain in real time, which ensures the authenticity and real-time performance of the service transaction data in the cloud manufacturing system, and thus ensures the authenticity and real-time performance of the service provider's *SC*.

Through the above calculation, we can see that *SC* has the following characteristics:

- The *SC* is proportional to the value of the service provided by the service provider in the cloud manufacturing system before, which makes the service provider pay more attention to the improvement of its own service value and quality;
- Among the services provided, the contribution of real-time services to *SC* is much higher than that of historical services. While measuring and evaluating the *SC* of service providers from a global perspective, the latest situation of enterprise development has been greatly considered.

- The user's evaluation of the use and experience of cloud services can affect the calculation of the service provider's *SC* to a considerable extent, so that the *SC* can comprehensively evaluate the enterprise cloud *SC* combining objective and subjective factors.

2. **Trust incentive mechanism**

   Trust is an indicator that describes the current trustworthiness of a member node in the cloud manufacturing service supply chain and the degree of fulfillment of obligations. It can quantify and characterize the trustworthiness of the member node and the activity of participating in the transaction and maintenance of the service chain, and is the specific value accumulated by the member node in the reward and punishment incentive mechanism of the service power consensus algorithm. Specifically, the trust degree is related to the specific behavior of the member node in the service chain, and is determined by the behavior of the member node in the system. Increase or decrease. The calculation formula of trust degree is as follows:

$$ Tr = \begin{cases} Tr + 2, & action{=}actionA \\ Tr + 5, & action{=}actionB \\ Tr + 3, & action{=}actionC \\ Tr - 30, & action{=}actionD \end{cases} \tag{10.5} $$

   Tr represents the trust degree of the member node. Action represents the action currently completed by the member node, and action is one of actionA, B, C, and D. actionA indicates that it agrees to participate in the consensus node election and successfully becomes a candidate node. actionB means to complete the packaging and production of transaction data in the blockchain and upload it to the chain. actionC means to complete the work of sorting candidate nodes according to their service capabilities and electing a new consensus node during the consensus node election. actionD indicates that the node broadcasts false transaction information, broadcasts false campaign credentials, or the consensus node fails to perform the relevant work of packaging transaction data and putting it on the chain [18].

   Among them, actionA is mainly driven by the consensus process, and is judged by state transition and broadcast of election credentials. actionB and actionC are judged by the generation of new blocks and the whole network broadcast, and are supervised by the nodes of the whole network. actionD is judged by all nodes of the network. Judgment is made by comparing the data in the public database and the local database, and broadcast to the whole network, and finally the consensus of the whole network is obtained.

3. **Service capability proof algorithm**

   For each node in the service chain, there are three states: normal node, candidate node, and leader node. The state transition diagram is shown in Fig. 10.6.

   - Nodes in the "normal node" state, when receiving the election notification, responds according to its own settings. It will become a "candidate

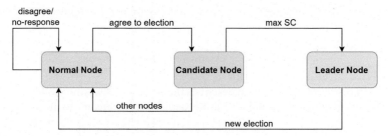

FIG. 10.6 State transition diagram of nodes in the service chain.

node" state if it agrees to participate in the consensus node election. Otherwise, it will give up this time and keep staying in the "ordinary node" state if it does not agree or does not reply.

- Nodes in the "candidate node" state will broadcast their campaign credentials and *SC* to each node. Other nodes will verify their *SC* based on the campaign credentials. The candidate node with the highest service capability will be converted to the "consensus node" state. Otherwise, it will turn to "normal node" state.

- Nodes in the "leader node" state are responsible for packaging and producing blocks of the received transaction information. Meanwhile, it sorts the candidate nodes during the election of the consensus node according to their *SC* to elect a new leader node. When a leader node is recommended but not elected, it will turn to the "normal node" state.

Fig. 10.7 shows the flow chart of the proposed service capability proof algorithm. In each period, the consensus node is responsible for the block production of five blocks, and will send a campaign notification to all nodes after the fourth block is completed. If the notification is answered with agreement, it will be converted into the "candidate node" state. If it refuses or does not respond, it means giving up the election of the consensus node, and it is still in the normal node state. Candidate nodes need to use the transaction data in the last *n* blocks as their own election credentials and broadcast to other nodes together with the service power calculated by the service power calculation formula.

When receiving the campaign credentials and service capabilities broadcasted by the candidate nodes, verification will be done. Meanwhile, the consensus nodes sort the candidate nodes according to the received service capabilities, elect the candidate node with the highest service capability as the new consensus node, and assign relevant information and transaction information to be packaged into the fifth block of this period. If it is not elected, it will be converted to the "ordinary node" state. The new consensus node is responsible for the block production of five blocks in the next period and the election of the new consensus node. When the candidate node

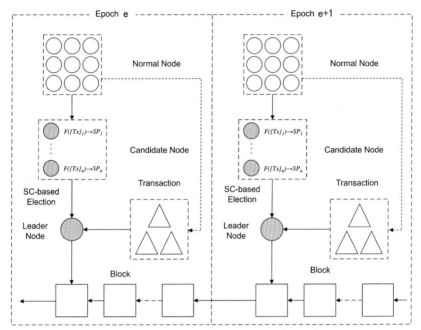

FIG. 10.7 Flow chart of the proposed service capability proof algorithm.

broadcasts false campaign credentials and service power, and the consensus node fails to complete the block production work or packages false transaction information, a certain amount of trust will be deducted as a penalty.

Each consensus node is responsible for the block production of five blocks for the following considerations: when the total number of transactions contained in the five blocks remains unchanged, splitting a large block into five small blocks reduces the size of the block. The amount of data in each block, on the one hand, can speed up the broadcast speed of the block, on the other hand, it can speed up the publication speed of transactions in the network, and further improve the security and credibility of the data in the network; when the five blocks contain When the total transaction volume of the blockchain network expands, on the one hand, it can improve the transaction processing speed of the blockchain network, on the other hand, it can reduce the number of consensuses, further reduce the energy consumption of the system, and improve the consensus efficiency.

The pseudocode of the consensus algorithm of the service capability proof consensus is shown in Algorithm 10.1. During the initialization process of the algorithm, the system generates a genesis block; evaluates all nodes in the system and gives them the corresponding initial trust degree Cr_i, and finally initializes the system parameters. Then, in each round of consensus election, consensus nodes, common nodes, and candidate nodes perform actions in sequence to perform their obligations, elect a new consensus node, perform state transitions, and finally reset the value of each counter.

Algorithm 10.1 Consensus algorithm of the service capability proof consensus

---

1: **Procedure:**
2: Generate genesis block
3: Initialization: Cr_i, T_c, M_bn
4: **While** *True*
5:   **If** $t \geq T_c$ or $bn \geq M\_bn$
5:     Leader: send *msg(Consensus)* to every node
6:     Nodes: respond to *msg(Consensus)*
7:       state transition according to response
8:     Leader: send *msg(Submit consensus credentials)* to every candidate node
9:     Candidate_Nodes: broadcast consensus credentials
10:         $Cr_{this\_node}{=}Cr_{this\_node} + 2$
11:     Leader: verification consensus credentials and calculate service power
12:     **If** consensus credentials not true
13:      $Cr_{this\_node}{=}Cr_{this\_node} - 30$
14:     **End if**
15:     Leader: select new Leader according to consensus credentials
16:      $Cr_{Leader\_node}{=}Cr_{Leader\_node} + 3$
17:     State transition
18:     $t = 0$
19:     $bn = 0$
20:   **End if**
21: **End while**

The pseudo-code of the transaction algorithm of the service capability proof consensus is shown in Algorithm 10.2. The algorithm initialization process is similar to that in the consensus algorithm, but different system parameters need to be initialized. During the transaction process, the nodes in the blockchain conduct transactions with each other, and broadcast the transaction information in the blockchain network, and the consensus nodes store the received transaction information in the transaction pool. When the transaction pool overflows or the timer expires, the consensus node encapsulates the transactions in the transaction pool into blocks, links the generated blocks to the blockchain, and broadcasts them to the entire network. Finally reset the value of each counter.

Algorithm 10.2 Transaction algorithm of the service capability proof consensus

---

1: **Procedure:**
2: Generate genesis block
3: Initialization: Cr_i, T_b, M_tx
4: **While** *True*
5: Every node in system: trading with other nodes
6:         broadcast *msg(transaction)*
7: Leader: receive *msg(transaction)* and put transaction into transaction pool
8: $tx = tx + 1$
9: **If** $t \geq T_b$ or $tx \geq M\_tx$
10: Leader: generate new block and add transactions from transaction pool to the block
11:      $Cr_{Leader\_node} = Cr_{Leader\_node} + 5$
12:      *clear(transaction pool)*
13:      broadcast *msg(block)*
14: $bn = bn + 1$
15: $t = 0$
16: $tx = 0$
17: **End if**
18: **End while**

Table 10.2 lists the symbols and descriptions of all parameters in Algorithms 10.1 and 10.2.

TABLE 10.2 Symbols and descriptions of all parameters in Algorithms 10.1 and 10.2.

Symbols	Descriptions
Cr_i	Trust value of the i-th node in the service chain
T_c	The maximum interval between two consensus processes
M_bn	The number of blocks each consensus node is responsible for generating
t	Current time of the timer
bn	The number of blocks generated by the current consensus node
T_b	Maximum time interval between two block generation processes
M_tx	The maximum number of transactions that can be recorded in each block
tx	The number of transactions recorded in the transaction pool
msg()	Indicates the content of information sent or broadcast
clear()	Clear function which clears the content recorded in the specified parameter

10.4 Intellectual property protection

Due to the particularity of the 3D printing industry, there is a high risk of intellectual property infringement in the cloud 3D printing platform. For example, an original 3D printed model can be easily copied and pasted to different places on the Internet like an electronic document, and the ease of such dissemination does not make it reasonable. For another example, 3D digital models of physical objects can be easily obtained by 3D scanning them. The original physical object can then be replicated using a 3D printer. If the cost of copying technology continues to decrease, the risk of plagiarism will increase significantly. Therefore, the corresponding intellectual property protection mechanism and digital license management mechanism are needed in the cloud 3D printing platform to ensure that the rights and interests of copyright owners are not infringed. Designers and creators of 3D printed models should be protected by intellectual property rights and paid reasonably. This is especially important. The risk of plagiarism will greatly increase if new business models emerge, such as when huge profits can be made by 3D printing spare parts. Therefore, with the continuous in-depth application of 3D printing technology in different aspects of the entire manufacturing process, we will face more challenges in authorizing access to product data, ensuring product delivery, distinguishing counterfeit products, and intellectual property protection. In the consumer sector, Article 53 of the German Copyright Act regulates the end user of additively manufactured components. When the number of copies is small (7 copies or less), copying for private use without the author's consent is permitted. These copies can also be sent to friends for free. However, consumers may not accept the return service of 3D printed models, as this is considered profitable and thus is an act of plagiarism [19].

Adding anti-counterfeiting signs to 3D printed products may be an effective way to address counterfeit and shoddy products, including visible product labels and invisible product labels. These product labels can be used during product processing as well as product recycling. Product labels can include some important information about the product and the manufacturer, such as the source of the product, the manufacturer. In addition, the product label should also have a symbol that can identify the uniqueness of the product. Visible product labels can be achieved by applying security labels or by holograms with a void effect. This visible product label is automatically destroyed when the seal is opened. RFID (Radio Frequency Identification) technology can be used to provide additional information about origin, supply chain or manufacturing parameters. To avoid signature fraud of RFID tags, additional tags can also be combined with high-resolution cloud-like print images. The human eye cannot see its subtleties. If a counterfeiter tries to imitate it, the picture loses precision and optical detail. Forgery can thus be detected using suitable reading devices. These images can also be attached to 3D printed products as direct markers. For example, the serial number is directly added to the surface of the product by means of stamping, laser, etc., so as to achieve its inseparability. It is also possible to use codes or pictures that are not verifiable to the counterfeiter.

In the cloud manufacturing platform 3D printing process chain, the preparation of the geometry, the determination of the process parameters or the manufacture of the model is usually done by other manufacturing service providers. Therefore, the cloud 3D printing platform must consider copyright issues. If the service provider prepares the geometric model for printing and subsequently creates the corresponding printing template through slicing software, he may have finally created a 3D printed digital model under the corresponding copyright laws. In this case, the resulting 3D model may not be reproduced and distributed without the permission of the copyright owner. Public availability also requires permission from the author or copyright holder. Additionally, the original product manufacturer may not be allowed to alter the prepared geometry. Therefore, the rules for legal boundary conditions must be clearly defined when contracting with service providers to create print templates. Printing a 3D model, on the other hand, means copying, as the 3D model is treated as a physical object. The 3D model itself is not changed by this, but only the representation. Therefore, the number of printed products does not matter, the first printed product is already a reproduction of the original. If the 3D printed model is passed on to a service provider for fabrication, he has no property rights in the protected work.

10.5   Conclusions

Security is an important aspect of cloud 3D printing platforms and services, and it is also an important guarantee for the wider and deeper application of cloud 3D printing. In this chapter, we discussed the security issues related to cloud 3D printing platforms, including data security of cloud 3D printing platforms, access control of cloud 3D printing devices, security of blockchain-based cloud 3D printing services, and the intellectual property protection problem. In terms of data security of cloud 3D printing, we mainly discussed the main data security problems (such as data leakage and illegal data access and tampering), and the main technical means to solve these problems (such as different data encryption and decryption technologies, violation identification and notification technology, etc.). In terms of access control of cloud 3D printing equipment, we mainly discussed the necessity of access control of 3D printing equipment in the cloud manufacturing platform, as well as the main technical means to achieve access control of 3D printing equipment, and gave specific method details. Then we discussed the blockchain-based cloud 3D printing service security technology, including the introduction of blockchain and its main applications in cloud manufacturing, blockchain-based cloud 3D printing service credit security, and its relationship with 3D printing Related intellectual product protection issues. The application of these blockchain technologies in the cloud manufacturing architecture strengthens the data sharing between different nodes in different regions in the cloud manufacturing platform. BC can make the data sharing in the cloud manufacturing system more secure and efficient, thereby improving the credit security between different entities in the cloud manufacturing system. Using blockchain technology to store and encrypt production data, transaction data and enterprise reputation data in the cloud manufacturing system can not only ensure the privacy and information security of enterprises in the cloud manufacturing system, but also improve the reputation of enterprises and cloud manufacturing platforms. Degree and safety. Blockchain technology can increase the dependence of manufacturing enterprises on cloud manufacturing platforms, break the trust barriers between different regions, different industries, and different enterprises, thereby solving the trust security problem of cloud manufacturing systems. Finally we discussed the intellectual property protection problem in the cloud 3D printing environment.

In future work, more attention should be paid to the research and practice of cloud 3D printing platform security, considering how to apply traditional network security technologies to specific cloud manufacturing 3D printing scenarios, and how to use advanced models such as blockchain to improve The credibility of the cloud manufacturing platform, how to increase the enthusiasm of the member nodes to participate in the maintenance of the cloud manufacturing blockchain system while maintaining the high degree of decentralization of the cloud manufacturing system.

References

[1] X. Liu, et al., Enhancing the security of cloud manufacturing by restricting resource access, J. Homel. Secur. Emerg. Manag. 11 (4) (2014) 533–554.

[2] Cloud Security Alliance: The Treacherous 12 - Cloud Computing Top Threats in, 2016. https://downloads.cloudsecurityalliance.org/assets/research/top-threats/Treacherous-12_Cloud-Computing_Top-Threats.pdf.

[3] C. Esposito, et al., Cloud manufacturing: security, privacy, and forensic concerns, IEEE Cloud Comput. 3 (4) (2016) 16–22.

[4] B. Buckholtz, R. Ihab, L. Wang, Remote equipment security in cloud manufacturing systems, Int. J. Manuf. Res. 11 (2) (2016) 126–143.

[5] S. Nakamoto, Bitcoin: A peer-to-peer electronic cash system, Decentralized Business Review, 2008, p. 21260.

[6] Z. Li, A.V. Barenji, G.Q. Huang, Toward a blockchain cloud manufacturing system as a peer to peer distributed network platform, Robot. Comput. Integr. Manuf. 54 (2018) 133–144.

[7] Z. Li, L. Liu, A.V. Barenji, et al., Cloud-based manufacturing blockchain: secure knowledge sharing for injection mould redesign, Procedia CIRP 72 (2018) 961–966.

[8] J. Lee, M. Azamfar, J. Singh, A blockchain enabled cyber-physical system architecture for industry 4.0 manufacturing systems, Manuf. Lett. 20 (2019) 34–39.

[9] B. Kaynak, S. Kaynak, Ö. Uygun, Cloud manufacturing architecture based on public blockchain technology, IEEE Access 8 (2019) 2163–2177.

[10] A.V. Barenji, Z. Li, W.M. Wang, Blockchain cloud manufacturing: shop floor and machine level, in: Smart SysTech 2018, European Conference on Smart Objects, Systems and Technologies. VDE, 2018, pp. 1–6.

[11] Y. Zhang, X. Xu, A. Liu, et al., Blockchain-based trust mechanism for iot-based smart manufacturing system, IEEE Trans. Comput. Soc. Syst. 6 (6) (2019) 1386–1394.

[12] C. Yu, L. Zhang, W. Zhao, S. Zhang, A blockchain-based service composition architecture in cloud manufacturing, Int. J. Comput. Integr. Manuf. 33 (7) (2020) 701–715.

[13] Q. Wang, C. Liu, B. Zhou, Trusted transaction method of manufacturing services based on blockchain, Comput. Integr. Manuf. Syst. 25 (12) (2019) 3247–3257.

[14] R. Dong, M. Yuan, Z. Zhou, Cloud manufacturing service transaction information recording technology based on blockchain, Comput. Technol. Dev. 29 (5) (2019) 97–101.

[15] A.V. Barenji, Z. Li, W.M. Wang, et al., Blockchain-based ubiquitous manufacturing: a secure and reliable cyber-physical system, Int. J. Prod. Res. 58 (7) (2019) 2200–2221.

[16] C. Dwork, M. Naor, Pricing via processing or combatting junk mail, in: Annual International Cryptology Conference, Springer, Berlin, Heidelberg, 1992, pp. 139–147.

[17] M. Castro, B. Liskov, Practical Byzantine fault tolerance, in: Proceedings of the Third USENIX Symposium on Operating Systems Design and Implementation (OSDI). New Orleans, LA, USA, February 22–25, 1999, pp. 173–186.

[18] Y. Zhang, L. Zhang, Y. Liu, X. Luo, Proof of service power: a blockchain consensus for cloud manufacturing, J. Manuf. Syst. 59 (2021) 1–11.

[19] M. Holland, J. Stjepandić, C. Nigischer, Intellectual property protection of 3D print supply chain with blockchain technology, in: 2018 IEEE International Conference on Engineering, Technology and Innovation (ICE/ITMC), 2018, pp. 1–8.

Chapter 11

Application demonstration of cloud 3D printing platform

Through extensive research and analysis of the shortcomings of existing commercial platforms, this book discusses in detail the customization-oriented cloud 3D printing in cloud manufacturing. This chapter introduces a cloud 3D printing platform prototype developed by the authors' lab based on the concepts, methods and technologies introduced in the previous chapters and the cloud 3D printing platform architecture in Chapter 3 (Fig. 3.3).

The platform effectively integrates the 3D printing production process, Users can get the whole process services of 3D printing including model design, model verification, order allocation, demand-supply matching and scheduling, task management, production process monitoring, after-sales and others on the platform. This chapter introduces how to realize a customized production process based on 3D printing through this platform.

11.1 Customized production based on cloud 3D printing platform

The portal of the cloud 3D printing platform is divided into six modules: home page, online customization, product mall, cloud service, consulting center and about us, as shown in Fig. 11.1. On the home page, users can understand the basic information of the platform, and at the same time can intuitively understand the process of using the platform for customized production. In the online customization module, users can customize the design and production of parts. The product mall displays 3D printing products and 3D model drawings to users, and users can directly purchase them online, as shown in Fig. 11.2. The cloud service section gathers service components with various functions, and users can learn the relevant information of their products in the customization process through various service components in this section, as shown in Fig. 11.3. The consulting center can regularly update the latest research progress in the field of 3D printing at home and abroad, as shown in Fig. 11.4.

Customized Production Through 3D Printing in Cloud Manufacturing
https://doi.org/10.1016/B978-0-12-823501-0.00010-9

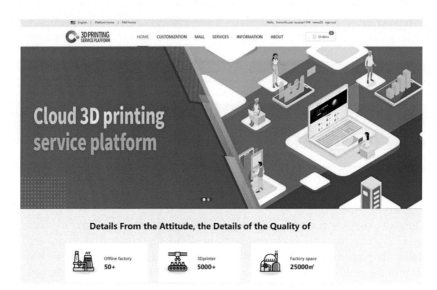

FIG. 11.1 The portal of the cloud 3D printing platform.

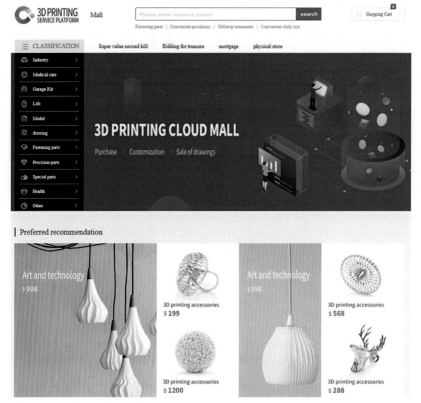

FIG. 11.2 Product mall module.

FIG. 11.3   Cloud service module.

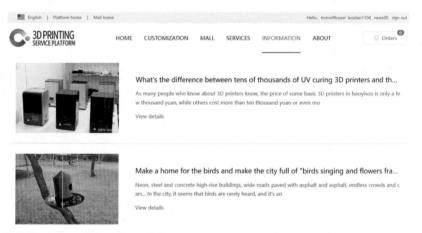

FIG. 11.4   Consulting center module.

The process of using this platform for customized design and production will be described in detail below. We take the customized storage box as a demonstration to introduce the whole process of the online customization section.

1. Requirements filling

The first thing for users to do is filling in the customization requirements according to the template after entering the online customization section, as

FIG. 11.5   Online customization—Fill in the customized requirements.

shown in Fig. 11.5, the name of the customized product is storage box, and the user can select the printing material in the printing material option. The future utilizations of custom products are divided into industrial and DIY, and the user will be assigned an industrial-level or desktop-level 3D printer according to the option platform. In the description module, the user can fill in the basic requirement information of the customized product. For example, the user requires an FDM 3D printer with the accuracy of 0.25 mm, and the printing material is yellow PLA. This description module is prepared for users with experience in 3D printing, and inexperienced users can also leave this field for designers to supplement relevant information. At the same time, the user can upload the 3D design model of the storage box or the photo of it, so that the designer can understand the user's customization requirements and accelerate the design process. Finally, in the permission option of the data usage, the platform's permission whether the platform can use the custom product for publicity, sales of parts drawings, sampling and analysis of design parameters, and other activities in the future is determined by users.

2. **Demand communication with designers**

    After the user submits the customization requirements online, the platform will announce these tasks to the designers who have joined the platform and have relevant expertise in the user's customized demand information. The designer receives the task in the platform and can communicate with the user face-to-face. After the user selects a suitable designer and connects with the designer, the front-end of the platform will display the basic information of the designer, including the job number, basic contact information, brief introduction to specialties, number of tasks received, and favorable rating, as shown in Fig. 11.6. The progress bar on the left will update the processing progress of the task.

3. **Confirmation of detailed requirements and payment**

    After the designer completes the geometric parameter design of the customized product, he needs to use simulation verification module of the design prototype to verify the credibility of the relevant parameters, and output the verification analysis report according to the credibility evaluation criteria, as shown in the Fig. 11.7. At the same time, on the front-end of the platform, users can see and download the designer's parametric design verification report. When both the user and the designer confirm online, the quotation information of the order will be displayed. Then, the next process will be performed after the user pays the payment, as shown in Fig. 11.8.

4. **3D renderings and parameter confirmation**

    After successful payment, users can browse and download the 3D renderings of their customized products online, and confirm the detailed design parameters, including materials, precision, dimensions and expected delivery date. After the confirmation online, the task will enter the next customization process, as shown in Fig. 11.9.

FIG. 11.6   Online customization—Designer take orders and ask requirements.

FIG. 11.7 Prototype parameter design validation report.

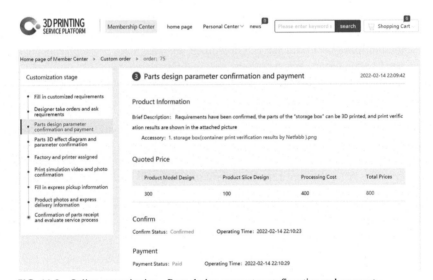

FIG. 11.8 Online customization—Parts design parameter confirmation and payment.

5. Factory and 3D printer automatic allocation

When the user confirms the detailed design parameters, the design procedure of the customized product is completed and the production procedure is entered. According to the supply and demand matching method described in Chapter 4, the customized products will be allocated to the appropriate 3D

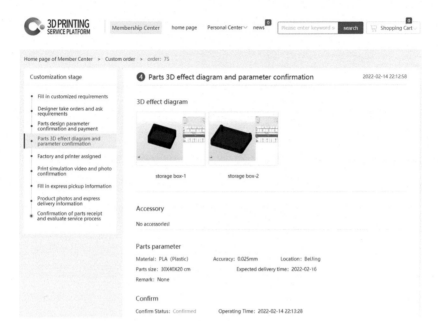

FIG. 11.9 Online customization—Parts 3D effect diagram and parameter confirmation.

printer based on the detailed design parameters of the customized parts and the detailed information of the 3D printer and its affiliated factories will be displayed, as shown in Fig. 11.10. At the same time, users can grasp the detailed information of all 3D printers connected to the platform through the print resource preview service component of the cloud service module, as shown in Fig. 11.11. The number and production capacity of printers owned by the manufacturing plants that join the platform through the component can also be found out in the platform. This information will be updated daily according to the daily schedule of the franchised factories, see Fig. 11.12.

6. Confirmation of 3D print simulation video and photo

After assigning a suitable 3D printer to the user, the technicians of the relevant franchise factories will design the printing parameters and present the simulation video and simulation 3D renderings through the simulation software to the user. The simulation video can indicate the actual printing effect of the 3D printer, and feedback the estimated printing time with high confidence to the user. Besides, the simulated 3D renderings can display detailed printing parameter design information. After the user confirms that these design parameters are correct, the task will enter the actual production process, as shown in Fig. 11.13.

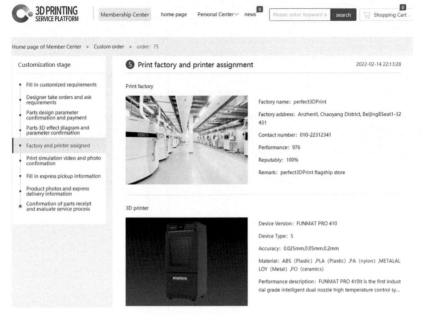

FIG. 11.10 Online customization—Print factory and printer assignment.

FIG. 11.11 Cloud services—Platform print resource preview.

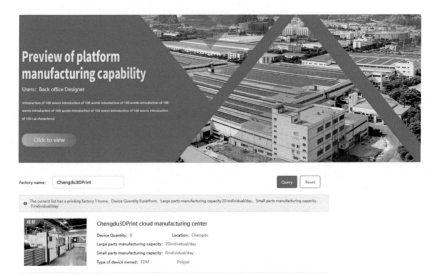

FIG. 11.12   Cloud services—Platform manufacturing capability preview.

7. **Consignee information filling**

   Users need to fill in their receipt information while waiting for the 3D printing to be finished. If the user has filled in the relevant receiving information before, the platform will automatically generate the receiving information.

8. **Information of photo and express**

   When the 3D printing production is completed, the technicians in the 3D printing factory will perform post-processing on the customized products, such as surface polishing and removal of printing supports, and send the final photos of the finished products to the user. If the user satisfies with the product, the factory will send the product to the user according to the receiving information. The express information can be found in the platform, as shown in Fig. 11.14.

9. **Confirm receipt and evaluation**

   Finally, the user receives the customized product, confirms the receipt and completes the follow-up evaluation form. Then, the user's receipt confirmation information, user evaluation results, evaluation opinions, and received physical photos will be displayed on the front-end of the platform. At the same time, the degree of disclosure of customized parts information in the user customization process will be reiterated in the order data privacy, as shown in Fig. 11.15.

FIG. 11.13 Online customization—Print simulation video and photo confirmation.

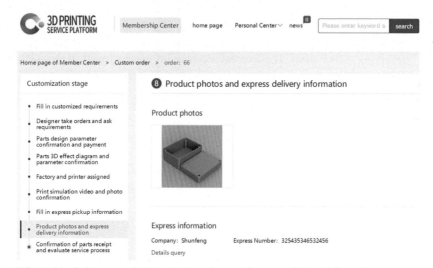

FIG. 11.14 Online customization—Product photos and express delivery information.

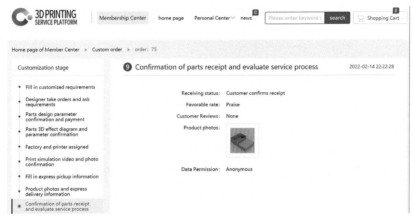

FIG. 11.15 Online customization—Confirmation of parts receipt and evaluate service process.

11.2 Conclusion

At present, the prototype system is only used for research in the laboratory, and many functions need to be improved and optimized, but it basically embodies the idea and implementation path of personalized manufacturing mode formed by the combination of 3D printing and cloud manufacturing. We hope it can provide reference for future research and practice.

Chapter 12

Conclusions and future work

Under the traditional manufacturing mode, there is an irreconcilable contradiction between product customization and manufacturing cost. The initial motivation of mass production is to reduce costs by improving efficiency. In traditional economic theory, large-scale production is the most important means to reduce costs. Customization will increase the cost, which has become a common sense.

As a revolutionary manufacturing technology, 3D printing has completely overturned the traditional manufacturing mode and provided an unprecedented powerful tool for realizing low-cost customized manufacturing. 3D printing dramatically simplifies the manufacturing process of products, and the advantage of a fully digitalized production process enables product design and manufacturing to be closely integrated. The article on the third industrial revolution published by the Economist in 2012 has made a good imagination for the future manufacturing industry based on the customized manufacturing ability of 3D printing [1]. But to truly realize this imagination, it is impossible to leave the support of cloud platform.

Cloud manufacturing, proposed in 2009, based on the philosophy of cloud computing, centralizes distributed manufacturing resources through the cloud service platform to form a virtual manufacturing enterprise with huge manufacturing capacity and unlimited expansion, which provides a revolutionary paradigm for the future of manufacturing industry.

After more than 10 years of development and practice, the concept of cloud manufacturing is being accepted by the industry. More and more companies are engaged in providing professional services of cloud manufacturing, such as CASICloud [2], RootCloud [3], KREATIZE [4], Fast Radius [5]. There are more and more startup cloud manufacturing companies all over the world [6]. Obviously, the manufacturing activities are increasingly dependent on platforms. Platforms unlock new sources of value creation and supply by using data-based tools to create community feedback loops [7]. Platform economy has become an important economic model, which is affecting every corner of human society.

Through the intelligent cloud service platform, the ability of cloud manufacturing can be perfectly combined with the characteristics of 3D printing

Customized Production Through 3D Printing in Cloud Manufacturing
https://doi.org/10.1016/B978-0-12-823501-0.00006-7
193

to share large-scale manufacturing resources and pave the way to realize the ideal customized manufacturing mode.

Dr. H. Lipson fully affirmed this model in his book *The new world of 3D printing* [8], and made optimistic predictions for its future. He pointed out: one future business model enabled by 3D printing and the new design technologies will be cloud manufacturing. Thousands of 3D printers with different characteristics would be like ants with factories. Each manufacturing company, alone, by itself, may be small. However, like billions of cell phones or ants with factories, the combined whole will be greater than the sum of its parts.

In this book, we firstly introduced some basic concepts including customized manufacturing, cloud manufacturing, as well as the possibility of combining them. We also discussed the irreconcilability between customization and cost under the traditional customized manufacturing mode, as well as the advantages of cloud 3D printing platforms which combine 3D printing and cloud manufacturing. Secondly, some key technologies for supporting the combination of 3D printing and cloud manufacturing were discussed in Chapters 4–10. Through the 3D printing model management and design technology, users can either draw sketches online to be matched with 3D models in the model library or generate 3D models by uploading photos of different angles of an object. Two 3D printer accessing methods were also introduced to connect the 3D printer to the cloud platform. The online monitoring technique of printing processes was then proposed to find out failures and faults earlier. Besides, the credibility evaluation methods of cloud 3D printing were discussed for both 3D models and printing services. Some important techniques of supply-demand matching and task scheduling in 3D printing platforms were also proposed to improve system efficiency. Studies of improving the efficiency of printing by effectively managing the printing process were then introduced such as parallel processing and task packaging. The security-related issues in cloud 3D printing were also discussed including data security, access control of online 3D printers, blockchain-enabled trust security, and intellectual property protection problems. Lastly, we introduced a cloud 3D printing platform prototype which was developed by the authors' team based on these methods and technologies introduced in the book. The customized production process and main functions of this platform were illustrated through the production process of a storage box.

It is evident that the combination of 3D printing and cloud manufacturing has become a stage to realize the revolutionary of manufacturing industry. The world has moved fast on this road. However, the technologies supporting cloud 3D printing are still on the way, and many operations on the platform are far from intelligence or even automation. Due to the openness, dynamics and uncertainty of the cloud platform, there are still many issues in carrying out extensive manufacturing business on a cloud platform. For example:

Customized 3D printing design: There is still a lot of work to be done for customized 3D printing design such as using advanced artificial intelligence technologies to achieve more accurate matching of hand-drawn sketches and existing 3D models, developing new transfer learning models to generate better 3D models based on images, and applying virtual prototype, digital twin and other related technologies to achieve more personalized collaborative 3D model designing in the cloud platform.

Customized manufacturing: There is also a lot of work to be done to realize more efficient, customized, and intelligent manufacturing of 3D printing such as reasonably and efficiently decomposing diverse 3D printing tasks, using most advanced AI techniques to quickly select the best 3D printing devices among a large number of services to execute specific tasks, to intelligently schedule printing resources to meet the needs of customers to the greatest extent, and to efficiently package multiple models, and using 5G technologies to improve the remote visual monitoring of 3D printers and printing processes.

Credibility and safety: In terms of credibility and safety, some open problems include how to effectively evaluate and verify the credibility of customers and manufacturing services on the cloud platform, how to effectively and reliably authenticate and evaluate users and services entering the platform, how to take advantage of blockchain and other related technologies to guarantee the safety, reliability, and efficiency of cloud transaction processes, and how to make full use of social resources to improve manufacturing capabilities, while protecting the core technical secrets from malicious use or disclosure.

Applications of emerging technologies: Emerging techniques are developing very rapidly in recent years including artificial intelligence, digital twin, cyber-physical system, virtual reality, quantum computing, etc. How to rightly use these technologies to enhance the cloud 3D printing platforms and solve long-standing problems is worthy to further study. Specifically, some open questions include how to apply artificial intelligence and big data methods to mine high-volume data, models, and knowledge on the platform, how to establish highly reliable and available digital twins of physical 3D printing resources to guarantee high-quality services, how to use digital twin and cyber-physical systems to make the manufacturing process more transparent so that users can participate better, how to use modeling and simulation, model-based system engineering and cloud VR to carry out cloud collaborative design, how to manage the virtual world or community formed by manufacturing users, and how to design and develop the new generation cloud platform based on quantum computers.

In addition, the popularity of cloud 3D printing also depends on the maturity of 3D printing technology, the reduction of 3D printing cost, the development of material science and technology, and how deeply the 3D printers are connected to the cloud platform. The current 3D printing technology is far from replacing the traditional manufacturing technology. For a long time, 3D printing and traditional manufacturing will coexist and make up for each other's shortcomings.

However, in the field of high customization requirements, 3D printing will play an increasingly important role.

There is still a long way to go to fully realize this revolutionary manufacturing mode. It requires the joint efforts of the whole society such as science and technology, law, culture and finance, and make use of all the latest technologies and ideas available to realize a more efficient and intelligent 3D printing cloud manufacturing platform to support customized, large-scale and low-cost manufacturing.

References

[1] I. Whadcock, The third industrial revolution, The Economist (2012). Apr 21st.
[2] http://www.indics.com.
[3] https://en.rootcloud.com.
[4] https://kreatize.com.
[5] https://www.fastradius.com.
[6] https://www.startus-insights.com/innovators-guide/discover-5-top-cloud-manufacturing-solutions-developed-by-startups.
[7] G.G. Parker, M.W.V. Alstyne, S.P. Choudary, Platform Revolution—How Networked Markets are Transforming the Economy and How to Make Them Work for you, W. W. Norton & Company, Inc, 2016.
[8] H. Lipson, M. Kurman, Fabricated: The New World of 3D Printing, John Wiley & Sons, 2013.

Index

Note: Page numbers followed by *f* indicate figures.

Printed in the United States
by Baker & Taylor Publisher Services